AF389176

Nanocelluloses

Nanocelluloses

Synthesis, Modification and Applications

Special Issue Editor

Elena Vismara

MDPI • Basel • Beijing • Wuhan • Barcelona • Belgrade • Manchester • Tokyo • Cluj • Tianjin

Special Issue Editor
Elena Vismara
Politecnico di Milano
Italy

Editorial Office
MDPI
St. Alban-Anlage 66
4052 Basel, Switzerland

This is a reprint of articles from the Special Issue published online in the open access journal *Nanomaterials* (ISSN 2079-4991) (available at: https://www.mdpi.com/journal/nanomaterials/special_issues/nano_cellulose).

For citation purposes, cite each article independently as indicated on the article page online and as indicated below:

LastName, A.A.; LastName, B.B.; LastName, C.C. Article Title. *Journal Name* **Year**, *Article Number*, Page Range.

ISBN 978-3-03928-784-0 (Pbk)
ISBN 978-3-03928-785-7 (PDF)

Cover image courtesy of Elena Vismara.

Contents

About the Special Issue Editor . . . . **vii**

Preface to "Nanocelluloses" . . . . **ix**

Elena Vismara, Andrea Bernardi, Chiara Bongio, Silvia Farè, Salvatore Pappalardo, Andrea Serafini, Loredano Pollegioni, Elena Rosini and Giangiacomo Torri
Bacterial Nanocellulose and Its Surface Modification by Glycidyl Methacrylate and Ethylene Glycol Dimethacrylate. Incorporation of Vancomycin and Ciprofloxacin
Reprinted from: *Nanomaterials* **2019**, *9*, 1668, doi:10.3390/nano9121668 . . . . **1**

Mingquan Zhang, Xiao Wu, Zhenhua Hu, Zhouyang Xiang, Tao Song and Fachuang Lu
A Highly Efficient and Durable Fluorescent Paper Produced from Bacterial Cellulose/Eu Complex and Cellulosic Fibers
Reprinted from: *Nanomaterials* **2019**, *9*, 1322, doi:10.3390/nano9091322 . . . . **23**

Davide Venturi, Alexander Chrysanthou, Benjamin Dhuiège, Karim Missoum and Marco Giacinti Baschetti
Arginine/Nanocellulose Membranes for Carbon Capture Applications
Reprinted from: *Nanomaterials* **2019**, *9*, 877, doi:10.3390/nano9060877 . . . . **33**

Andrew Colburn, Ronald J. Vogler, Aum Patel, Mariah Bezold, John Craven, Chunqing Liu and Dibakar Bhattacharyya
Composite Membranes Derived from Cellulose and Lignin Sulfonate for Selective Separations and Antifouling Aspects
Reprinted from: *Nanomaterials* **2019**, *9*, 867, doi:10.3390/nano9060867 . . . . **51**

J. Elliott Sanders, Yousoo Han, Todd S. Rushing and Douglas J. Gardner
Electrospinning of Cellulose Nanocrystal-Filled Poly (Vinyl Alcohol) Solutions: Material Property Assessment
Reprinted from: *Nanomaterials* **2019**, *9*, 805, doi:10.3390/nano9050805 . . . . **69**

Zhouyang Xiang, Jie Zhang, Qingguo Liu, Yong Chen, Jun Li and Fachuang Lu
Improved Dispersion of Bacterial Cellulose Fibers for the Reinforcement of Paper Made from Recycled Fibers
Reprinted from: *Nanomaterials* **2019**, *9*, 58, doi:10.3390/nano9010058 . . . . **85**

Inese Filipova, Velta Fridrihsone, Ugis Cabulis and Agris Berzins
Synthesis of Nanofibrillated Cellulose by Combined Ammonium Persulphate Treatment with Ultrasound and Mechanical Processing
Reprinted from: *Nanomaterials* **2018**, *8*, 640, doi:10.3390/nano8090640 . . . . **99**

Selestina Gorgieva and Janja Trček
Bacterial Cellulose: Production, Modification and Perspectives in Biomedical Applications
Reprinted from: *Nanomaterials* **2019**, *9*, 1352, doi:10.3390/nano9101352 . . . . **111**

About the Special Issue Editor

Elena Vismara graduated in Chemistry in 1978 at the Università degli studi di Milano, Italy. She began her career as Assistant Professor at the Politecnico di Milano in 1978. In 1992, she became Associate Professor at the Politecnico di Milano. She is Head of the Applied Organic Chemistry Laboratory (AOCL). The main AOCL activities are organic and inorganic chemical preparations, reaction workups, separation and purification by the main organic and inorganic chemistry techniques, structural and morphological characterizations, and basic analytical chemistry. Elena Vismara's team has been working in multiple directions of basic and applied research, in more or less sophisticated fields. Many of Elena Vismara's projects have been financed by public or private funds. The team has national and international collaborations, the most relevant and ongoing with Ronzoni Institute of Milan, Italy (www.ronzoni.it); Technion, Haifa, Israel (https://www.technion.ac.il/en/technion-israel-institute-of-technology/); Sachim Company (http://www.sachim.it/); and Emodial company (http://www.emodial.it/).

She is the author of around 80 papers and inventor of about 15 patents to date.

Selected publications

1. Nanocellulose: preparation and modification.

Vismara, Elena et al. Bacterial nanocellulose and its surface modification by glycidyl methacrylate and ethylene glycol dimethacrylate. incorporation of vancomycin and ciprofloxacin. *Nanomaterials* (2019), 9(12), 1668–1690.

2. Preparation of hybrid organic–inorganic nanoparticles for biomedical applications.

Vismara, Elena et al. Albumin and hyaluronic acid-coated superparamagnetic iron oxide nanoparticles loaded with paclitaxel for biomedical applications. *Molecules* (2017), 22(7), 1030/1–1030/25.

3. Recovery of cellulose from natural and industrial waste for the synthesis of artificial fibers.

Santanocito Adriana, Vismara Elena. Production of textile from citrus fruit. PCT Int. Appl. (2015), WO 2015018711 A1 20150212.

4. Polymeric materials from natural and synthetic fibers for sanitary and environmental applications.

Vismara, Elena et al. Polyvinyl acetate processing wastewater treatment using combined Fenton's reagent and fungal consortium: Application of central composite design for conditions optimization. *Journal of Hazardous Materials* (2018), 358, 243–255.

5. Synthesis of carbohydrate derivatives with antimetastatic properties.

Borsig, Lubor; Vismara, Elena et al. Sulfated hexasaccharides attenuate metastasis by inhibition of P-selectin and heparanase. *Neoplasia (Ann Arbor, MI, United States)* (2011), 13(5), 445–452

6. Preparation of low molecular weight heparin.

Vismara, Elena; Mascellani, Giuseppe; et al. Torri, Giangiacomo. Low-molecular-weight heparin from Cu2+ and Fe2+ Fenton type depolymerisation processes. *Thrombosis and Haemostasis* (2010), 103(3), 613–622.

Selected patents

1. Polyurethane-hydrogel dressing comprising silver nanoparticles. Pecorari, Federico; Vismara, Elena, PCT Int. Appl. (2018), WO 2018055203 A1 20180329.

2. Polyurethane-hydrogel composition comprising chlorhexidine.   Vismara, Elena; Pecorari, Federico, PCT Int. Appl. (2018), WO 2018055200 A2 20180329.
3. Polyethylene net or fabric grafted with a PVP hydrogel for the absorption and release of pyrethroids (2019) US2019000078 (A1), Vismara Elena, Starace Giuseppe, Arrigoni Paolo.

Preface to "Nanocelluloses"

Nanocelluloses: Synthesis, Modification and Applications

Elena Vismara

Department of Chemistry, Materials and Chemical Engineering "G. Natta", Politecnico di Milano, via Mancinelli 7, 20131 Milano, Italy. elena.vismara@polimi.it; Tel.: +39-0223993088

Nanocelluloses (NCs), namely cellulose-based materials with peculiar physicochemical properties, have appeared as a novel material suitable for a wide range of specific applications that are quite distinct from those of cellulose. Since 2011, two reviews have highlighted both the NC structure and applications of the new developed materials. NC structure, properties, and nanocomposites have been critically reviewed [1], and NCs have been defined as a new family of nature-based materials [2]. Recently, a handbook of nanocellulose and cellulose nanocomposites has been published [3]. Nanocellulose-based biomaterials have been significantly investigated for biomedical applications due to their excellent physical and biological properties, such as biocompatibility and low cytotoxicity [4]. Nanocellulose has also been connected to the topic of sustainability [5]. A nanocellulose that has been studied in recent years and that is prepared by biotechnology is bacterial nanocellulose (BNC), a nanofibrillar polymer produced by several species of bacterial [6]. Furthermore, from a chemical point, BNC is chemically identical with plant cellulose but is free of byproducts like lignin, pectin, and hemicelluloses, featuring a unique reticulate network of fine fibers, and these are the physicochemical peculiarities that make BNC different from other nanocellulose materials [7]. This Special Issue includes 7 articles and a review, covering BNC [8, 9], bioactive BNC [10], and a review of BNC and its biomedical applications [11]. Two articles concern nanocellulose membranes used for carbon capture applications [12] and composite membranes derived from cellulose and lignin sulfonate used for selective separations and in antifouling [13]. Finally, electrospinning was applied to nanocrystal-filled poly(vinyl alcohol) solutions [14] and nanofibrillated cellulose was obtained by combining ammonium persulfate oxidation with common mechanical treatment [15].

1. Moon, R.J.; Martini, A.; Nairn, J.; Simonsen, J.; Youngblood, J. Cellulose nanomaterials review: structure, properties and nanocomposites. *Chem. Soc. Rev.* **2011**, *40*, 3941–3994.

2. Klemm, D.; Kramer, F.; Moritz, S.; Lindström, T.; Ankerfors, M.; Gray, D.; Dorris, A. Nanocelluloses: A new family of nature-based materials. *Angew. Chem. Int. Ed. Engl.* **2011**, *50*, 5438–5466.

3. Kargarzadeh, H.; Ahmad, I.; Thomas, S.; Dufresne, A. *Handbook of Nanocellulose and Cellulose Nanocomposites*; Wiley-VCH Verlag GmbH & Co. KGaA: Weinheim, Germany, 2017; 2 Volumes, pp. 1-443, 1-849.

4. Jorfi, M.; Foster, E.J. Recent advances in nanocellulose for biomedical applications. *J. Appl. Polym. Sci.* **2015**, *132*, 41719/1–41719/19.

5. Lee, K-Y. *Nanocellulose and Sustainability: Production, Properties, Applications, and Case Studies*, 1st ed.; CRC Press, Taylor & Francis Group: Boca Raton, FL, USA, 2018; pp. 1–284.

6. Gama, M.; Gatenholm, P.; Klemm, D. *Bacterial NanoCellulose: A Sophisticated Multifunctional Material*; CRC Press/Balkema: Leiden, The Netherlands, 2013; pp. 1–272.

7. Bacakova, L.; Pajorova, J.; Bacakova, M.; Skogberg, A.; Kallio, P.; Kolarova, K.; Svorcik, V. Versatile Application of Nanocellulose: From Industry to Skin Tissue Engineering and Wound Healing. *Nanomaterials-Basel* **2019**, *9*, 164–203.

8. Zhang, M.; Wu, X.; Hu, Z.; Xiang, Z.; Song, T.; Lu, F. A Highly Efficient and Durable Fluorescent Paper Produced from Bacterial Cellulose/Eu Complex and Cellulosic Fibers. *Nanomaterials* **2019**, *9*, 1322.

9. Xiang, Z.; Zhang, J.; Liu, Q.; Chen, Y.; Li, J.; Lu, F. Improved Dispersion of Bacterial Cellulose Fibers for the Reinforcement of Paper Made from Recycled Fibers. *Nanomaterials* **2019**, *9*, 58.

10. Vismara, E.; Bernardi, A.; Bongio, C.; Farè, S; Pappalardo, S.; Serafini, A.; Pollegioni, L.; Rosini. E.; Torri, G. Bacterial nanocellulose and its surface grafting and cross-linking by glycidyl methacrylate and ethylene glycol dimethacrylate. Preparation, characterization and incorporation of vancomycin and ciprofloxacin. *Nanomaterials* **2019**, *9*, 1668.

11. Gorgieva, S.; Trček, J. Bacterial Cellulose: Production, Modification and Perspectives in Biomedical Applications. *Nanomaterials* **2019**, *9*, 1352.

12. Venturi, D.; Chrysanthou, A.; Dhuiège, B.; Missoumand, K.; Giacinti Baschetti, M. Arginine/Nanocellulose Membranes for Carbon Capture Applications. *Nanomaterials* **2019**, *9*, 877.

13. Colburn, A.; Vogler, R.J.; Patel, A.; Bezold, M.; Craven, J.; Liu, C.; Bhattacharyya, D. Composite Membranes Derived from Cellulose and Lignin Sulfonate for Selective Separations and Antifouling Aspects. *Nanomaterials* **2019**, *9*, 867.

14. Sanders, J.E.; Han, Y.; Rushing T.S.; Gardner, D.J. Electrospinning of Cellulose Nanocrystal-Filled Poly(Vinyl Alcohol) Solutions: Material Property Assessment. *Nanomaterials* **2019**, *9*, 805.

15. Filipova, I.; Fridrihsone, V.; Cabulis, U.; Berzins, A. Synthesis of Nanofibrillated Cellulose by Combined Ammonium Persulphate Treatment with Ultrasound and Mechanical Processing. *Nanomaterials* **2018**, *8*, 640.

Elena Vismara
Special Issue Editor

 *nanomaterials*

Article

Bacterial Nanocellulose and Its Surface Modification by Glycidyl Methacrylate and Ethylene Glycol Dimethacrylate. Incorporation of Vancomycin and Ciprofloxacin

Elena Vismara [1],*, Andrea Bernardi [1], Chiara Bongio [1], Silvia Farè [1], Salvatore Pappalardo [1], Andrea Serafini [1], Loredano Pollegioni [2], Elena Rosini [2] and Giangiacomo Torri [3]

[1] Department of Chemistry, Materials and Chemical Engineering "Giulio Natta", Politecnico di Milano, via Mancinelli 7, 20131 Milano, Italy; andrea.bernardi@polimi.it (A.B.); chiara.bongio@polimi.it (C.B.); silvia.fare@polimi.it (S.F.); salvatore.pappalardo@polimi.it (S.P.); andrea.serafini@polimi.it (A.S.)

[2] Department of Biotechnology and Life Sciences, Università degli Studi dell'Insubria, via J.H. Dunant 3, 21100 Varese, Italy; loredano.pollegioni@uninsubria.it (L.P.); elena.rosini@uninsubria.it (E.R.)

[3] Istituto Scientifico di Chimica e Biochimica "Giuliana Ronzoni", via Giuseppe Colombo 81, 20133 Milano, Italy; torri@ronzoni.it

* Correspondence: elena.vismara@polimi.it

Received: 22 September 2019; Accepted: 20 November 2019; Published: 22 November 2019

Abstract: Among nanocelluloses, bacterial nanocellulose (BNC) has proven to be a promising candidate in a range of biomedical applications, from topical wound dressings to tissue-engineering scaffolds. Chemical modifications and incorporation of bioactive molecules have been obtained, further increasing the potential of BNC. This study describes the incorporation of vancomycin and ciprofloxacin in BNC and in modified BNC to afford bioactive BNCs suitable for topical wound dressings and tissue-engineering scaffolds. BNC was modified by grafting glycidylmethacrylate (GMA) and further cross-linking with ethylene glycol dimethacrylate (EGDMA) with the formation of stable C–C bonds through a radical Fenton-type process that involves generation of cellulose carbon centred radicals scavenged by methacrylate structures. The average molar substitution degree MS (MS = methacrylate residue per glucose unit, measured by Fourier transform infrared (FT–IR) analysis) can be modulated in a large range from 0.1 up to 3. BNC-GMA, BNC-EGDMA and BNC-GMA-EGDMA maintain the hydrogel status until MS reaches the value of 1. The mechanical stress resistance increase of BNC-GMA and BNC-GMA-EGDMA of MS around 0.8 with respect to BNC suggests that they can be preferred to BNC for tissue-engineering scaffolds in cases where the resistance plays a crucial role. BNC, BNC-GMA, BNC-EGDMA and BNC-GMA-EGDMA were loaded with vancomycin (VC) and ciprofloxacin (CP) and submitted to release experiments. BNC-GMA-EGDMA of high substitution degree (0.7–1) hold up to 50 percentage of the loaded vancomycin and ciprofloxacin amount, suggesting that they can be further investigated for long-term antimicrobial activity. Furthermore, they were not colonized by *Staphylococcus aureus* (S.A.) and *Klebsiella pneumonia* (K.P.). Grafting and cross-linking BNC modification emerges from our results as a good choice to improve the BNC potential in biomedical applications like topical wound dressings and tissue-engineering scaffolds.

Keywords: bacterial nanocellulose; methacrylate; Fenton reagent; cross-linking; vancomycin; ciprofloxacin; bioactive bacterial nanocellulose

1. Introduction

Nanocelluloses (NCs), namely cellulose-based materials with peculiar physicochemical properties, appear as a new option offering a wide range of specific applications quite different from cellulose. Nanocellulose-based biomaterials have been significantly investigated for biomedical applications due to their excellent physical and biological properties like biocompatibility and low cytotoxicity [1].

A nanocellulose that has been studied in recent years is bacterial nanocellulose (BNC), a nanofibrilar polymer produced by several species of bacteria [2]. BNC is a hydrogel containing 1% of nanocellulose. It cannot be ignored by researchers interested in nanocellulose due to its unique properties, such as chemical purity, biocompatibility, inertness and non-toxicity, biofunctionality and hypoallergenicity, good mechanical strength, high absorbency, and the possibility of forming any shape and size. Due to its properties, the study and use of BNC are focused on biomedicine. Tissue engineering has been associated with BNC mostly because of its low cytotoxicity, high porosity, biocompatibility, and non-resorbability [1]. For soft-tissue implants and cartilage replacements, the fibrillar network of BNC offers high tensile mechanical properties and a hydrogel-like behaviour as BNC interacts with the surrounding water medium. However, despite recent advances, there are still many challenges to overcome before the full potential of BNC can be completely realized as a choice of material in tissue-engineering applications. Different physical modification, i.e., modification to change porosities, crystallinities, and fibre densities, and chemical modifications, i.e., modification of the chemical structures and functionalities, have been investigated to enhance BNC properties suitable for its applications in tissue-engineering applications [3].

In the recent review on nanocellulose as a natural source for groundbreaking applications in materials science, BNC's future has been widely discussed [4]. The key problems seem to be BNC bioreactor designs and implant production costs. Economic considerations related to different applications must be taken into account: a high-price implant material, for example, in a first artificial heart bypass, can be more competitive than a Nata de Coco-based food thickener obtained in mass production. The conclusion of authors of reference [4] is that nowadays there are no standard answers regarding BNC production costs. Noteworthy, some BNC-based medical products are on the market and are Food and Drug Administration (FDA) approved and European Community (CE) -certified.

Besides the efforts dedicated to the advancement in production technologies and the estimation of various pitfalls associated with BNC production, a more scientific question is intriguing, if BNC and its composites with other polymers/nanoparticles can become the material of preference in current regenerative medicine, starting from biomedical applications of BNC and functionalized BNC in drug delivery, tissue engineering and antimicrobial wound healing [5]. A very recent paper stating the BNC haemocompatibility favours developing BNC biomedical applications [6]. Furthermore, from a chemical point of view, BNC is chemically identical with plant cellulose but is free of byproducts like lignin, pectin, and hemicelluloses, featuring a unique reticulate network of fine fibres and these are physicochemical peculiarities that make the difference between BNC and other nanocellulose materials [7]. BNC has been already been proposed as a skin-tissue repair material in vivo to replace conventional gauze dressings. According to the results of many studies in the field of wound healing, BNC has been shown to be a superior candidate for conventional wound-dressing materials [8]. For tissue engineering and wound healing, the problems of infection and inflammation have to be carefully considered as BNC does not possess any antimicrobial or anti-inflammatory characteristics. BNC can be endowed with antimicrobial or anti-inflammatory activities by drug loading. Drug loading in BNC-based carriers includes physical absorption/adsorption and chemical conjugation [9]. BNC hydrogels can be fabricated using as low as 1% nanocelluloses, which may benefit from a large water content, that allows for the drugs to be loaded via diffusion and adsorption; however, the downside of this process is the prolonged loading time, which is not practical for clinical or industrial purposes. The loading time is a minimum 24 h and can reach 48 h. As the amount of the loaded drug is evaluated by the release experiments it is difficult to measure the real adsorbed amount.

Anyway, the most common loading method involves the immersion of the BNC delivery system in the drug solution. For example, BNC has been saturated with the antibiotic fusidic acid [10] and tetracycline hydrochloride has been loaded on bacterial cellulose composite membranes for controlled release [11]. BNC can be addressed for the delivery of a broad range of bioactive cargos [9]. The antiseptic octenidine has been used to provide active wound dressings based on bacterial nanocellulose [12]. The behaviour of BNC membranes has been studied as systems for topical delivery of lidocaine, an anaesthetic drug with high solubility in water in the form of hydrochloride and commonly used in surgery and topical application [13].

Since 2009 a review described the design and applications of biodegradable cellulose-based hydrogels and suggested that cellulose-based hydrogels, including BNC, could be ideal platforms for the design of scaffolding biomaterials in the field of tissue engineering and regenerative medicine [14]. The advantage of BNC is that it is obtained by a strain directly in the hydrogel status. Hydrogel-based devices have been developed for controlled drug delivery.

The aim of this study is to provide bioactive BNCs in the form of hydrogel, suitable for tissue engineering, regenerative medicine and wound dressing. We focus our attention on modified BNCs, in particular grafted and cross-linked BNC, made bioactive by loading drugs. The study concerns also the loading of BNC as it is and the comparison between BNC and modified BNC. The modifications were pursued by grafting glycidylmethacrylate (GMA) and cross-linking with ethylene glycol dimethacrylate (EGDMA). GMA-grafted and EGDMA-cross-linked BNCs of peculiar chemical and mechanical properties were obtained by a surface modification that maintained the original network structure of BNC. As concerns this last aspect, our study can be positioned in a relationship with BNC nanocomposites processing techniques that allow the incorporation of functional nanoreinforcements, nanofillers and additional phases without disturbing the original network structure of BNC [15]. The choice of GMA was justified by previous results that stated that GMA appendages endow cellulose with new properties due to the surface GMA network formation that acts like a molecular sponge [16]. GMA insertion is a radical-based process that was investigated in term of mechanism, detailed in all the synthetic aspects and huge supported by different characterisation techniques. Cellulose maintained its peculiar properties, as the methodology to insert GMA did not induce a coating by avoiding its polymerisation. GMA-modified cellulose was defined as a multitasking material as it found applications in quite different fields [16]. The adsorption and release properties of antibiotics is strictly related to the aim of this paper [17]. GMA-grafted fabrics loaded with vancomycin and other antibiotics were developed for topic antibacterial activity [18]. For BNC modification, GMA was also associated to EGDMA, a very efficient cross-linking agent already tested on nanocellulose [19] and on imprinted hydrogel prepared from *N,N*-dimethylacrylamide and tris(trimethylsiloxy)sililpropyl methacrylate as the main components, methacrylic acid as functional monomer [20]. The idea of using EGDMA comes not only from the need to reinforce BNC, but also from the hope that GMA and EGDMA can act in synergy in holding drugs. BNC and GMA-grafted and EGDMA-cross-linked BNCs were made bioactive by loading the antibiotics vancomycin and ciprofloxacin. Vancomycin is a glycopeptide antibiotic used in the treatment of infections caused by Gram-positive bacteria, such as *Staphylococcus aureus*. Vancomycin has been found many applications for bone regeneration. Notably, it does not interfere with normal in vivo bone regeneration [21]. It has also been selected as an antibiotic drug for enhanced bone regeneration [22]. Finally vancomycin has been used as polyhydroxy antibiotic for bio-inspired cross-linking and matrix–drug interactions for advanced wound dressings with long-term antimicrobial activity [23]. The broad-spectrum antibiotic ciprofloxacin has been used as model drug for BCN-cyclodextrin adduct to form drug-nanocarrier systems [24].

2. Materials and Methods

Hydrogen peroxide (H_2O_2) solution 30% (*w/w*), glycidyl methacrylate (GMA), ethylene glycol dimethacrylate (EGDM), iron sulfate heptahydrate ($FeSO_4 \cdot 7H_2O$), sodium hydroxide (NaOH), hydrochloric acid (HCl, 36.0–38.0% in water) and corn step liquor were purchased from Sigma

Aldrich (St. Louis, MO, USA). *Gluconacetobacter xylinus* (ATCC 10245) was purchased from the American Type Culture Collection (ATCC, Manassas, VA, USA).

2.1. Synthesis of Bacterial Nanocellulose (BNC)

BNC was obtained by the use of the bacterium *Gluconacetobacter xylinus* (ATCC 10245). The microorganism was inoculated in 5 mL ATCC medium 1 (5 g/L yeast extract, 3 g/L peptone, 25 g/L mannitol) and grown at 26 °C under static conditions for 48 h. An aliquot of 1 mL of this culture was transferred in a 250 mL Erlenmeyer flask containing 40 mL of corn steep liquor (CSL)-glucose medium (26 g/L CSL, 20 g/L glucose, 2.7 g/L Na_2HPO_4, 1.15 g/L citric acid monohydrate). The inoculated flasks were incubated at 30 °C for 144 h under static conditions. The BNC disk was removed from the medium and washed exhaustively with distilled water. Overnight treatment with 70 mL NaOH 1M followed by washing with water until neutral pH bleached the BNC sample. The final weight was 2.56 g.

2.2. Grafting of BNC

BNC was put in a three-neck round-bottom flask equipped with a thermometer and a condenser, containing 25 mL water for 1 g of BNC, reaction scale and reagent amount being reported in Tables 1 and 2. The suspension was maintained under mechanic stirring at 80 °C for 30 min. H_2O_2 and $FeSO_4$ (Fenton reagent) were added and left activating cellulose for 1 h at 80 °C. GMA was dropped in the flask and left to react for 30 min. BNC was filtered using a Gooch funnel and exhaustively washed with water at room temperature and with water at 45 °C to remove residual inorganic and organic reagents. The washed BNC was stored at 8 °C.

Table 1. Fenton-type glycidylmethacrylate (GMA) grafting. Relationship between reagent molar ratio and bacterial nanocellulose-GMA methacrylate residue per glucose unit (BNC-GMA MS).

Entry	Glucose (mmol)	$FeSO_4$ (mmol)	H_2O_2 (mmol)	GMA (mmol)	MS
1	1	6.50×10^{-1}	800	220	> 3
2	1	3.25×10^{-1}	400	110	2.8
3	1	1.63×10^{-1}	200	55	1.8
4	1	8.1×10^{-2}	100	30	0.8
5	1	4.0×10^{-2}	50	14	≈0.0
6 [1]	1	3.20×10^{-1}	920	610	0

[1] Reaction temperature 60 °C.

Table 2. Fenton-type GMA grafting in function of the reagent amount.

Entry [1]	BNC/Glucose	$FeSO_4$ [2]	H_2O_2 [3]	GMA	MS
1	2.1 g (1 mol)	0.10 mL (0.04 mol)	0.66 mL (49.5 mol)	0.24 mL (13.7 mol)	0.12–0.32
2	2.1 g (1 mol)	0.16 mL (0.06 mol)	1.00 mL (74.36 mol)	0.35 mL (20.53 mol)	≈0.25
3	2.1 (1 mol)	0.21 (0.08 mol)	1.32 (99.13 mol)	0.47 (27.37 mol)	0.3–0.7

[1] In bracket, the molar ratios normalized on 1 mmol of glucose unit. [2] 0.05 M $FeSO_4 \cdot 7H_2O$ solution. [3] H_2O_2 solution 30% (*w/w*).

2.3. Grafting and Cross-Linking of BNC

BNC was put in a three-neck round-bottom flask equipped with a thermometer and a condenser, containing 50 mL water for 2.2 g of BNC, reaction scale and reagent amount being reported in Table 3. The suspension was maintained under mechanical stirring at 80 °C for 30 min. H_2O_2 and $FeSO_4$ (Fenton reagent) and left activating cellulose for 1 h at 80 °C. GMA and EGDMA were simultaneously dropped in the flask and left reacting for 30 min. BNC was filtered using a Gooch funnel and exhaustively washed with water at room temperature and with water at 45 °C to remove residual inorganic and organic reagents. The washed BNC was stored at 8 °C.

Table 3. Fenton-type GMA grafting and ethylene glycol dimethacrylate (EGDMA) cross-linking in function of the reagent amount.

Entry [1]	BNC/Glucose	FeSO$_4$	H$_2$O$_2$	GMA	EGDMA	MS
1	2.2 g (1 mol)	0.16 mL (0.06 mol)	1.0 mL (74.36 mol)	0.26 mL (14.37 mol)	0.156 mL (6.2 mol)	0.6–0.74
2 [2]	2.2 g (1 mol)	0.16 mL (0.06 mol)	1.0 mL (74.36 mol)	0.26 mL (14.37 mol)	0.156 mL (6.2 mol)	≈1
3	0.534 (1 mol)	0.04 mL (0.06 mol)	0.308 mL (74.36 mol)	0 0	0.128 mL (20.53)	0.55

[1] In bracket, the molar ratios normalized on 1 mmol of glucose unit. [2] Reaction times 1 h.

2.4. Fourier Transform Infrared (FT–IR) Spectroscopy Analysis

The solid phase Fourier transform infrared (FT–IR) spectra of the powdered sample, obtained by BNC lyophilization and mixed with infrared-grade KBr, were generated using an ALPHA spectrometer (Bruker, Bremen, Germany). Data were analyzed using OPUS software, version 7.0 (Bruker, Bremen, Germany). The acquisition of the spectra were performed in the range 4000–400 cm^{-1}. The estimation of average molar substitution ratio (MS) was obtained with the method used in our previous works, Equation (1) [16,25]:

$$\text{MS}_{\text{FT–IR}} = \frac{\text{area ester (manual band integration)}}{\text{area cellulose (range } 780 - 465 \text{ cm}^{-1} \text{ integration)}} \tag{1}$$

At least two tests were carried out to assure reproducibility and accuracy for the integration calculations.

2.5. Solid-State Cross-Polarization Magic-Angle Spinning (CP-MAS) ^{13}C Nuclear Magnetic Resonance (NMR)

The ^{13}C nuclear magnetic resonance (NMR) analysis was performed by using a dipolar decoupling cross-polarization magic-angle spinning (DD-CP-MAS) technique with a Bruker Avance 300 spectrometer (75.47 MHz, Bruker, Milano, Italy). Concerning data acquisition parameters: the repetition time (D1) was equal to 8 s, while the contact time and the spin rate were 2 ms and 1000 Hz, respectively. 2 K scans were collected to obtain good quality spectra. The samples were positioned in a zirconium rotor (diameter: 4 mm; height: 21 mm). Tetramethylsilane was the reference substance for the chemical shifts. Benzene was used as secondary reference standard.

The crystallinity index (Cr. I %) of BNC materials was evaluated by means of Equation (2) [16,25]. A corresponds to integrals of the C-4 peaks at 86–92 ppm (crystalline) and B to the integrals of the C-4 peaks at 80–86 ppm (amorphous).

$$\text{Cr.I } (\%) = \frac{A}{A + B} \times 100 \tag{2}$$

2.6. Scanning Electron Microscopy (SEM) Experiments

Scanning electron microscopy (SEM) micrographs of the film surfaces were obtained on a Zeiss EVO 50 microscope (Zeiss, New York, NY, USA), equipped with a LaB$_6$ electron gun and operating at 15 kV. SEM was used to measure BNC fibre diameters and to study their morphologies. Prior to the SEM observation, a few nanometers gold film was deposited onto the sample surface in order to avoid a charging effect during SEM testing. The BNC fibers diameters statistical distributions were obtained analyzing the SEM micrographs with ImageJ (version 1.52r, NIH, National Institutes of Health, Bethesda, MD, USA) software [26], collecting more than 300 diameter sizes for each specimen.

2.7. Mechanical Properties

The structure and the properties of BNC strongly depend on the choice of the cellulose forming bacterial strain. The tensile strength changes significantly, too. For *Gluconacetobacter xylinus* ATCC 10245 the tensile strength is reported 0.184 mPa ([2], p. 185). Tensile strength and elongation to failure of BNC samples of 2 cm length, 0.5 cm width and 0.5 thickness were measured with a Dynamic Mechanical Analyzer (model DMA Q800, TA instruments, Milano, Italy). Samples were air dried for 2 h, with a water reduction content of the 50%. The compression reduced the water content of the 80%.

2.8. Adsorption and Release Experiments

Aqueous solutions of vancomycin (VC) and ciprofloxacin (CP), having a concentration 2×10^{-3} M were prepared. The CP solution was at pH = 3 with HCl 0.02 N to solubilize CP. BNCs (1.5 g, solid nanocellulose content 15 mg) were put into contact with the drug solution (25 mL). The flask was shaken in a thermostatic bath Julabo at 25 °C, (model SW22, 100 rpm, JULABO GmbH, Seelbach, Germany) for 24 h to get the equilibrium. Adsorption solutions were analyzed at different times with an ultraviolet (UV) spectrophotometer. Loaded BNCs were submitted to release experiments by removing the BNC from the adsorption solution and transferring it into a flask with deionized pure water (25 mL). The amount of loaded drug release was evaluated by submitting at different times the release solution to UV spectrophotometry.

2.9. Ultraviolet (UV) Spectroscopy Analysis

The amount of adsorbed/released molecules over time was monitored through quantitative ultraviolet–visible (UV–Vis) analysis (UV-spectrophotometer JASCO V-650; Jasco Europe, Cremella LC, Italy), Spectra Manager^{TM} software (version II, Jasco Europe, Cremella LC, Italy). Several aliquots of drug solution were collected at different times and submitted to UV acquisition. The acquisition wavelength range was set from 500 to 240 nm. The absorbance values of VC peak at λ_{max} = 280 nm and CP peak at λ_{max} = 272 nm were converted in concentration by using previously created calibration curves.

2.10. Antibacterial Activity Tests

The tests were run according to UNIEN ISO 20645:2005 by Centrocot (www.centrocot.it). *Staphylococcus aureus* (ATCC 6538 LOT: DSM 799-0415) and *Klebsiella pneumoniae* (ATCC 4352 LOT: DSM 789-0513) were cultured in Triptone and Soya medium for 24 h at 37 °C. The suspensions obtained were diluted to 10^8 UFC/mL. 1 mL samples (10^8 UFC/mL) were added to 150 mL of Triptone Soya Agar (LOT: Oxoid 1837431) at 45 °C to afford culture medium. A BNC sample of size 25 ± 5 mm was incubated in 5 mL of the culture medium for 24 h at 37 °C. The test was repeated on four equal BNC samples. Reference material was 100% cotton fabric according to International Organization for Standardization (ISO) 105/F02:2009.

3. Results

The synthetic procedures were first developed and optimised. Both modified BNCs and BNC were exhaustively characterised. Finally modified BNCs and BNC as it is were loaded with the drugs. The loaded BNCs were submitted to release studies and to in vitro antibacterial activity tests.

3.1. Grafting and Cross-Linking of BNC

This process provides BNC-GMA, BNC-EGDMA and BNC-GMA-EGDMA, as shown by Figure 1.

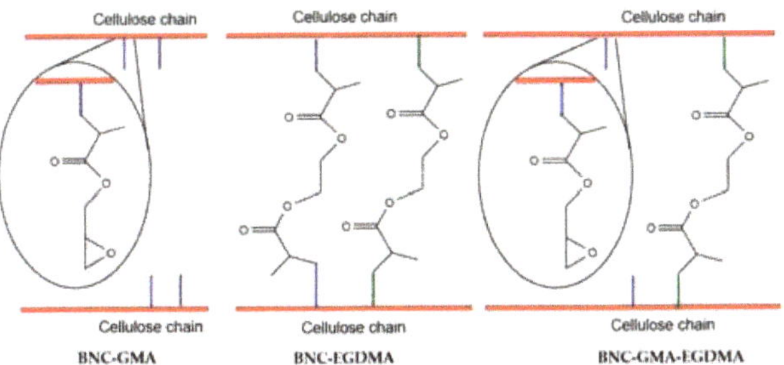

Figure 1. BNC-GMA (left); BNC-EGDMA (centre); BNC-GMA-EGDMA (right).

BNC was modified by the radical process that involves hydrogen peroxide and iron salt (Fenton reagent) generation of cellulose carbon centred radicals scavenged by glycidyl methacrylate (GMA) and by ethylene glycol dimethacrylate (EGDMA), see Scheme 1 [16,25].

Scheme 1. Grafting and cross-linking mechanism. Radical generation and trapping.

As the hydrogen abstraction is unselective due to the high reactivity of OH, the grafting and cross-linking occurs randomly on the cellulose chain. For convenience, only C4 radical was put in evidence in the Scheme 1. GMA scavenging results in branching cellulose, while EGDMA scavenging results in cross-linking cellulose. To quantify the scavenging yields the average molar substitution degree (MS) was defined as the number of GMA/EGDMA residue per glucose unit and measured by FT–IR spectrum as detailed in materials and methods paragraph. In agreement with the random functionalization MS is an average measure. Nevertheless it is important to point out that with this approach cellulose is not coated by a methacrylate polymer, as shown by FT–IR and NMR spectra, but only functionalized.

Figure 2 is the picture of BNC and grafted and cross-linked BNC. The functionalization reduces the BNC transparency that appears whiter.

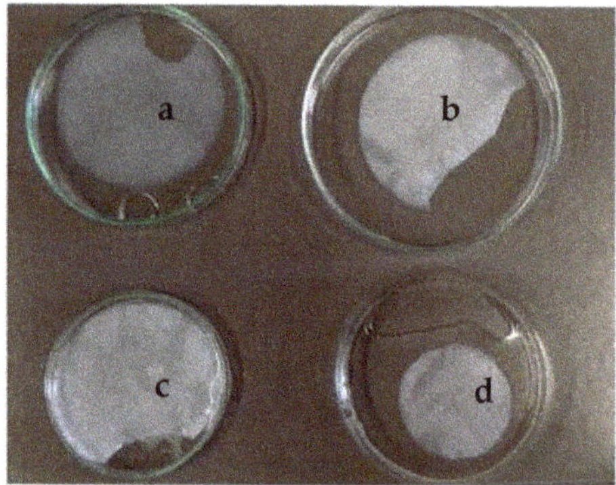

Figure 2. BNC (**a**); BNC-GMA MS = 0.25 (**b**); BNC-GMA-EGDMA MS = 0.8 (**c**); BNC-EGDMA MS = 0.55 (**d**).

3.1.1. Glycidylmethacrylate (GMA) Grafting

Table 1 reports the GMA grafting results on 390 mg of BNC containing 3.9 mg of nanocellulose (2.4×10^{-2} mmol of glucose unit) in function of the reagents amount. Every experiment was repeated at least three times. On this scale, the grafting reproducibility and homogeneity is very good. The molar ratios are normalized on 1 mmol of glucose unit (162 Da).

The BNC FT–IR spectrum can be seen in Figure 3.

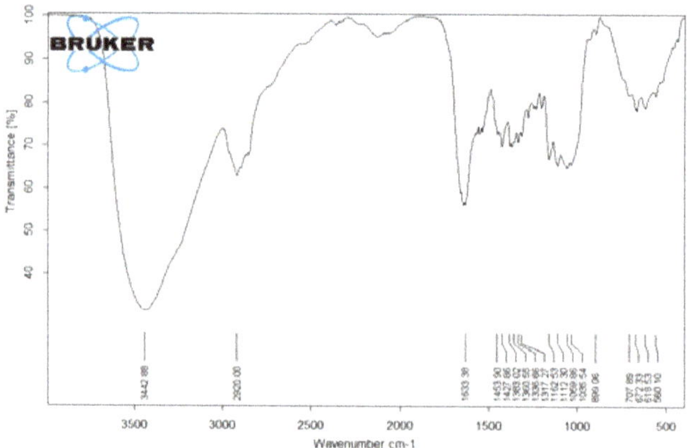

Figure 3. BNC Fourier transform infrared (FT–IR) spectrum: cellulose profile.

The BNC-GMA MS = 0.8 FT–IR spectrum (Figure 4) confirms that the GMA grafting occurred by the appearance of the bands of the glycidyl ester peak signal at 1728.57 cm^{-1}, of the C–H stretching signal of the epoxide ring signal at 3050–3000 cm^{-1} and of the C–O–C stretching signal of the epoxide ring at 1270 and 906.58 cm^{-1}. In the spectrum of Figure 4 we put in evidence the epoxide bands (*), the band of the glycidyl ester (*) and the bands typical of the cellulose in the range 780–465 cm^{-1} (C–O–C, C–C–O, and C–C–H deformation and stretching vibrations), whose integrations ratio in Equation (1) is defined MS.

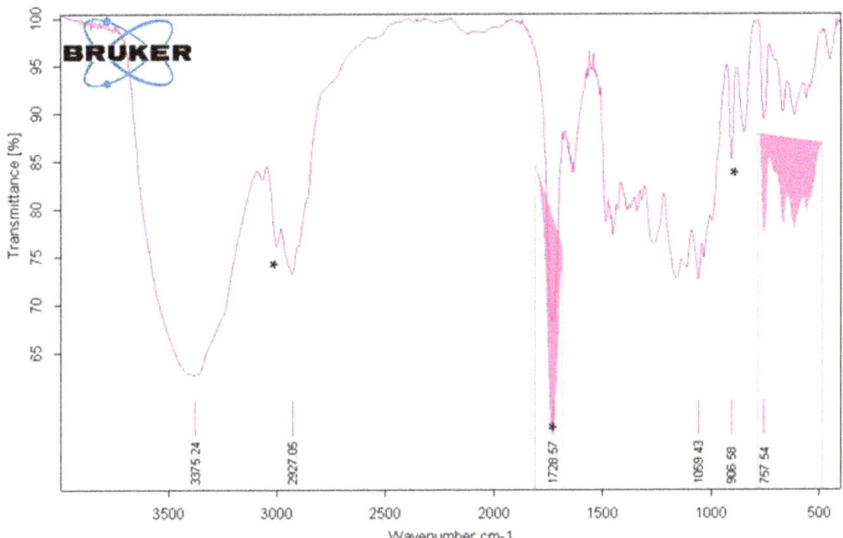

Figure 4. BNC-GMA MS = 0.8 FT–IR spectrum: cellulose profile *plus* GMA appendages (*).

In agreement with the reagents amount, see entries 3 and 4, Table 1, the BNC-GMA in entry 4, MS = 0.8, is less functionalized than the BNC-GMA in entry 3, MS = 1.8, as measured from their FT–IR spectra shown in Figures 4 and 5, respectively.

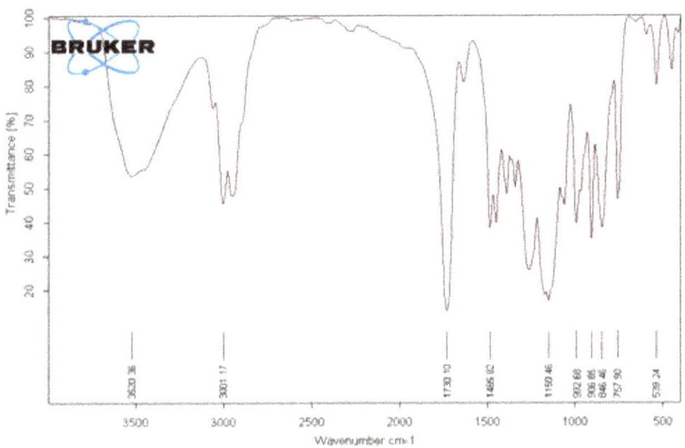

Figure 5. BNC-GMA MS = 1.8 FT–IR spectrum: cellulose profile *plus* GMA appendages.

Table 2 reports the GMA grafting on 2.1 g of BNC in function of the reagents amount.

When the reaction scale increases to more than 1 g, the grafting occurs with an acceptable reproducibility, being every experiment repeated at least three times, but it is not so easy to have homogeneous samples, as shown by the macroscopic aspect of the samples. Some zones appear more transparent than others, especially at low MS. Thus, every GMA BNC grafted sample was carefully mapped by FT–IR analysis that confirms the non-homogeneity. For convenience we report one spectrum, i.e., in Figure 6 the FT–IR spectrum of the less functionalized zone of MS = 0.12. In Table 2 the MS range calculated by FT–IR mapping is reported.

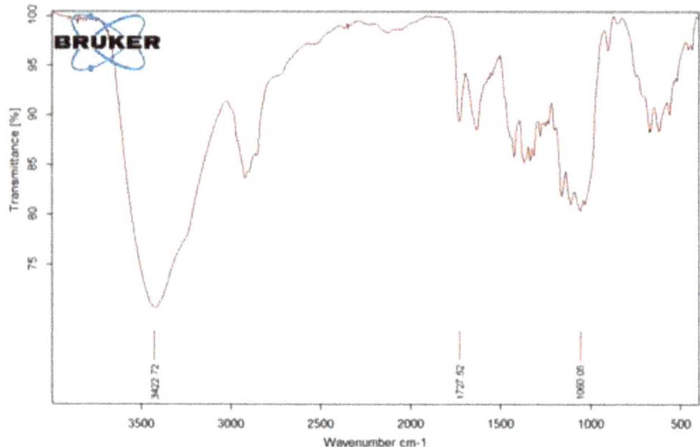

Figure 6. BNC-GMA MS = 0.12 FT–IR spectrum: cellulose profile *plus* GMA appendages.

Data in Tables 1 and 2 show that the GMA grafting can be modulated in a large range of MS. Until MS around 1, the hydrogel state was maintained. The water retention is little influenced by the functionalization. The percentage of cellulose measured by sample lyofilization changes from 1% for the starting BNC to 2–3% for MS around 0.6 and reaches 5% for MS around 1. The situation dramatically changed for highly functionalized BNC (MS > 2) that became solid and the water content was reduced to the half.

CP-MAS ^{13}C NMR spectra not only confirm the occurrence of the GMA grafting, but allow measurement of the crystallinity, see Equation (2). Figures 7 and 8 show the BNC solid-state CP-MAS spectrum, the second one with the measure of the integrals used to calculate the Cr. I % = 69.

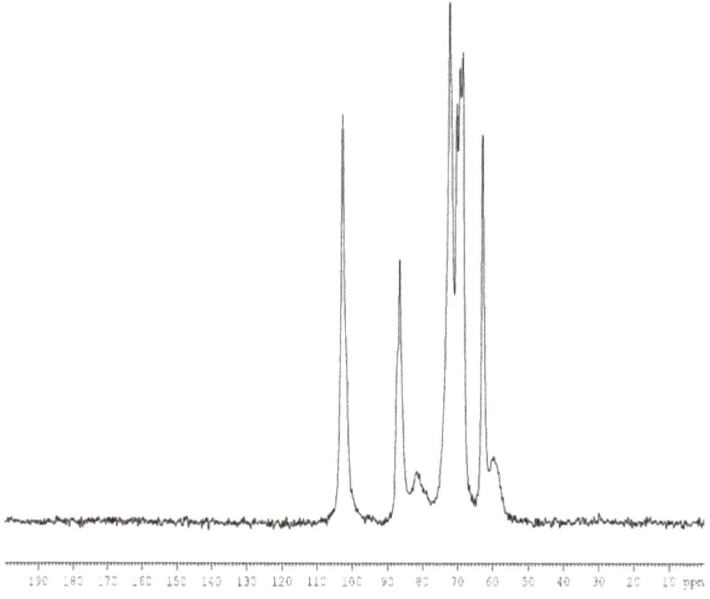

Figure 7. Cross-polarization magic-angle spinning (CP-MAS) ^{13}C nuclear magnetic resonance (NMR) of BNC: cellulose signals.

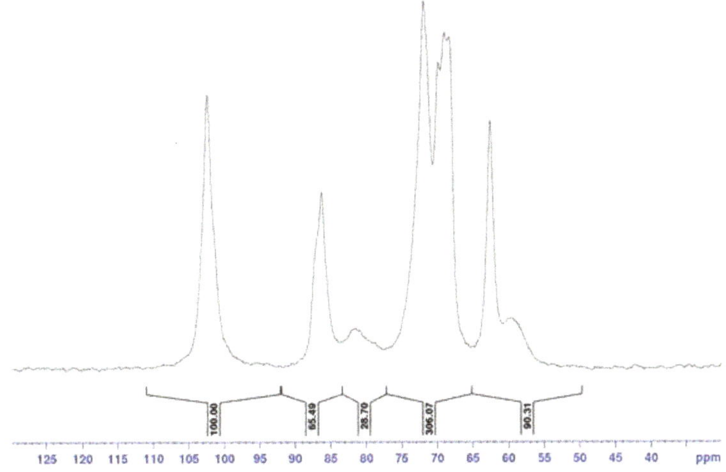

Figure 8. CP-MAS ^{13}C NMR of BNC-integration.

Figure 9 shows the CP-MAS ^{13}C NMR of BNC-GMA MS = 0.68 and its integration to measure Cr. I % = 69.

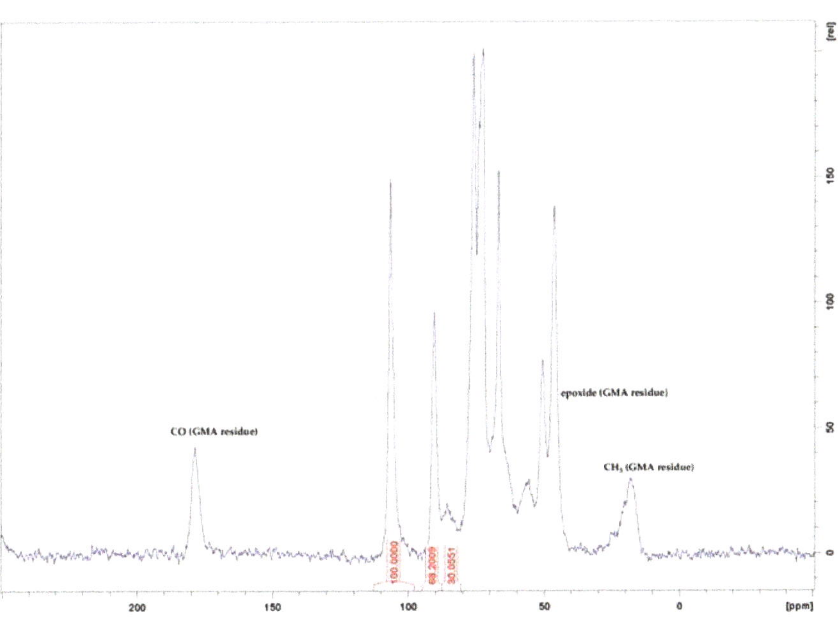

Figure 9. CP-MAS ^{13}C NMR of BNC-GMA MS = 0.68 -integration: cellulose signals *plus* GMA signals.

Figure 10 shows that even highly functionalized BNC-GMA maintains the crystallinity. The conclusion is that GMA grafting does not modify the crystallinity of BNC as it is.

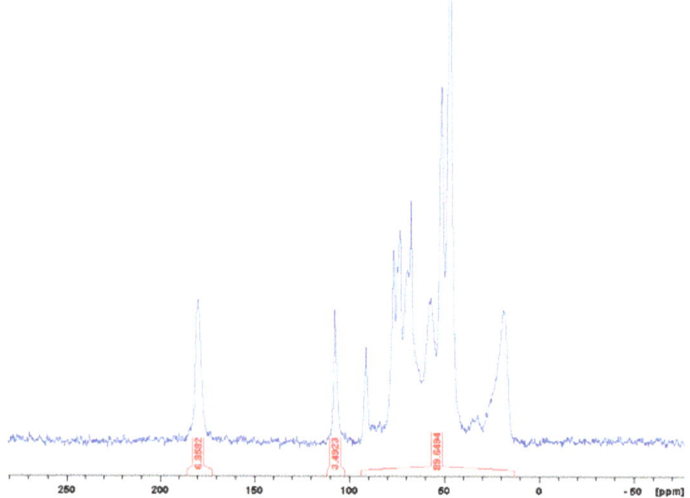

Figure 10. CP-MAS ^{13}C NMR of BNC-GMA (MS >> 1)-integration Cr. I % ≈ 69: cellulose signals *plus* GMA signals.

3.1.2. GMA Grafting and Ethylene Glycol Dimethacrylate (EGDMA) Cross-Linking

The EGDMA cross-linking conditions were decided on the base of the results obtained for GMA grafting, as summarized in Table 3. GMA and EGDMA were put in the reactions at the same time in the molar ratio 2:1. EGDMA was also used alone, see entry 3. The quantitative data show that grafting/cross-linking seems more efficient than the only grafting, see entries 1 and 3, Table 3, compared with entry 2, Table 1. In entry 2, MS = 1 was reached by increasing the scavenging time from 30 min to 1 h.

BNC-GMA-EGDMA FT–IR spectra at different MS are reported in Figures 11 and 12. As expected, spectra are quite similar to BNC-GMA spectra, due to the presence of the ester groups of GMA and EGDMA that are overlapped. They show also the epoxide group signals. The spectrum of BNC-EGDMA reported in Figure 13 shows only the ester group signal.

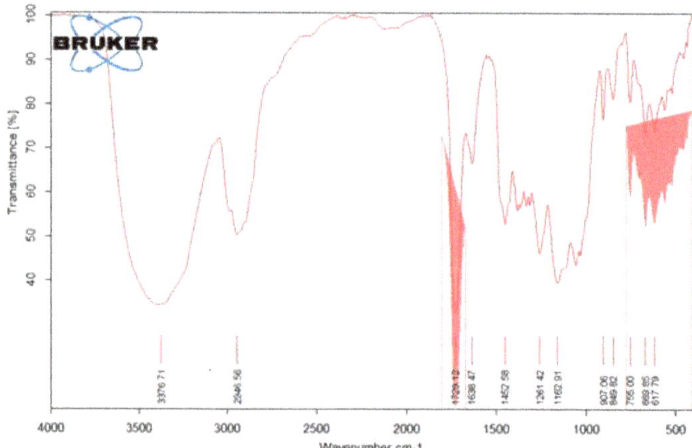

Figure 11. BNC-GMA-EGDMA MS = 0.6 FT–IR spectrum. GMA and EGDMA ester groups signal and cellulose reference signals in evidence.

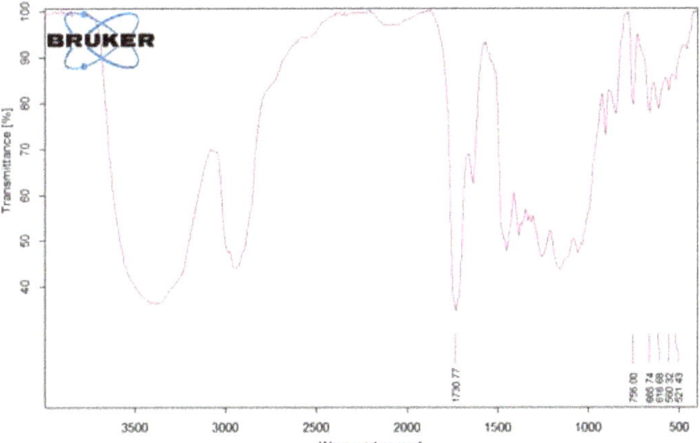

Figure 12. BNC-GMA-EGDMA MS = 1 FT–IR spectrum.

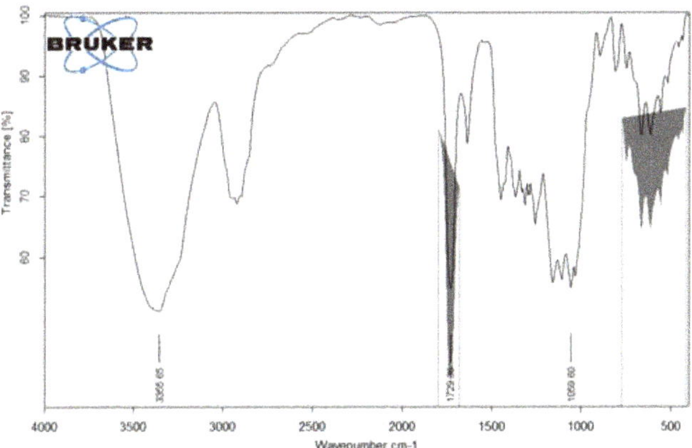

Figure 13. BNC-EGMA MS = 0.5 FT–IR spectrum. EGDMA ester group signal and cellulose reference signals in evidence.

CP-MAS ^{13}C NMR high functionalized BNC-GMA-EGMA (MS >> 1) spectrum reported in Figure 14 confirms the occurrence of the GMA and EGDMA grafting by comparison with the BNC-GMA spectrum as the methacrylate residues are quite similar, but not equal, see the difference with BNC-GMA spectrum for 10–30 peaks ($\underline{C}H_3$ aliphatic signals) and the difference in the range 60–80 peaks ($\underline{C}$–O signals). Also for high functionalized BNC-GMA-EGMA the crystallinity is maintained respect to the starting BNC.

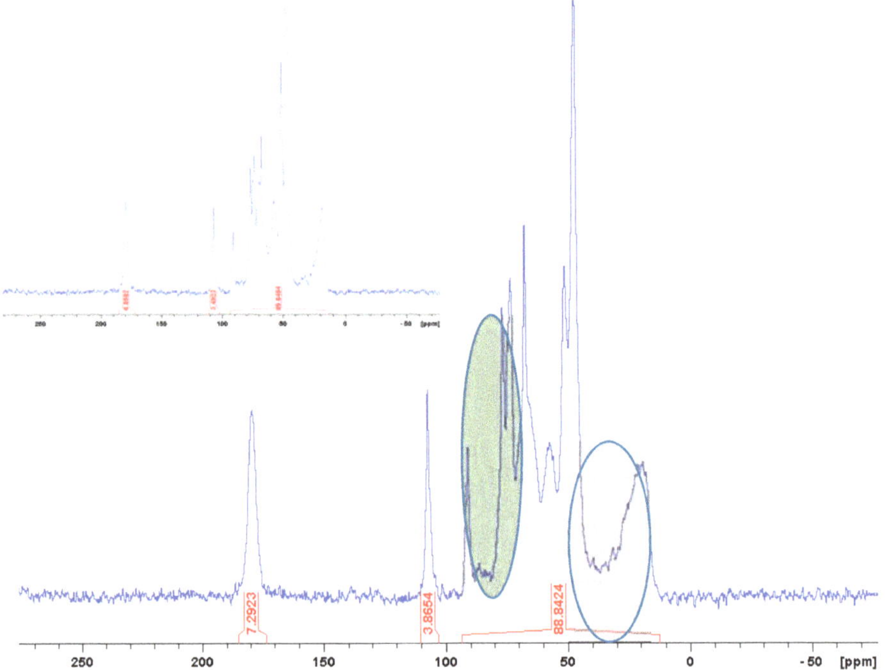

Figure 14. CP-MAS ^{13}C NMR of BNC-GMA-EGDMA (MS >> 1)-integration Cr. I % ≈ 69. Comparison with BNC-GMA spectrum (top left) of quite the same MS. Grey background $\underline{C}$H$_3$, green background $\underline{C}$-O.

3.2. SEM Analysis

The SEM morphological analysis and the measurements of the fibres diameters were conducted on the following samples: BNC as it is, BNC-GMA MS = 0.64, and BNC-GMA-EGDMA MS = 0.55, see Figure 15.

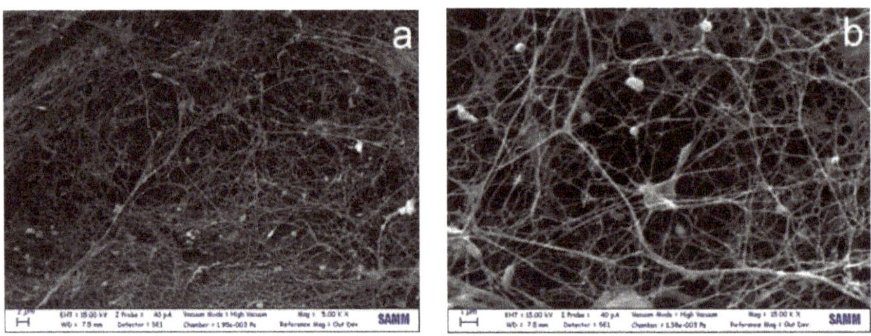

Figure 15. *Cont.*

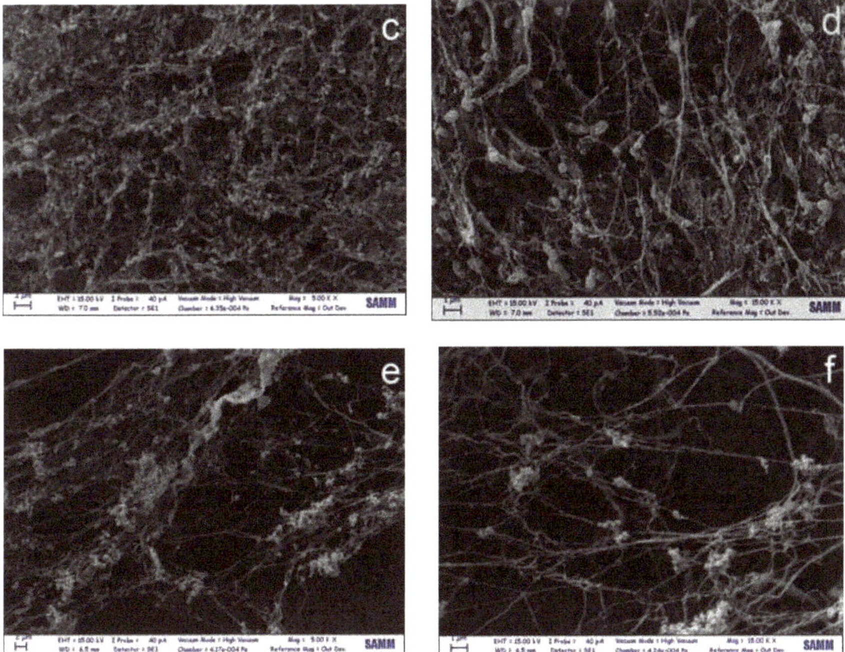

Figure 15. Scanning electron microscopy (SEM) micrographs of bacterial nanocellulose: (**a**,**b**) BNC as it is; (**c**,**d**) BNC-GMA (0.64 MS); (**e**,**f**) BNC-GMA-EGDMA (0.55 MS).

The BNC as it is specimen shows the presence of a dense network of fibres with a mean diameter of 84 ± 21 nm (see Table 4). It is also noted the presence of few pseudo-spherical structure (with dimension around 1 μm) on the network.

Table 4. Results from the dimensional analysis of BNC fibres displayed in the SEM micrographs.

Sample	BNC	BNC-GMA 0.64	BNC-GMA-EGDMA 0.55
mean fibres diameters	84 ± 21 nm	57 ± 12 nm	55 ± 11 nm

In the BNC-GMA MS = 0.64 sample the fibres are thinner and more packed respect to the BNC as it is sample. The fibre mean diameter is 57 ± 12 nm (see Table 4). Also in this sample, the pseudo-spherical structure with dimensions of 1 μm are present and they are more numerous respect to the BNC as it is sample.

Lastly, the BNC-GMA-EGDMA MS = 0.55 sample shows the presence of thin cellulose fibres with a mean diameter of 55 ± 11 nm (see Table 4) similar to the BNC-GMA MS = 0.64 specimen. However, in this case the fibres network is less dense respect to both the samples mentioned above. Moreover, the pseudo-spherical structures with 1 μm diameter are not present, and in their place few bunches composed by spherical particles with dimension of around 200 nm can be noted on the fibre network.

3.3. Stress–Strain Graphic

From the Figure 16 it appears that both BNC grafting and cross-linking modification enhance the mechanical resistance respect to the BNC as it is. Our strategy was not to dry the samples but only to reduce the water content with the aim of testing samples as similar as possible to the real BNC. Tested samples were left in the air for 2 h and lost the 50% of the water. In 5 h they can lose most of the water. The GMA grafting alone with MS = 0.64 significantly reinforced the fibres in the hydrogel. The grafting/cross-linking with a quite similar MS had a higher effect. The best BNC in term of resistance is the BNC-GMA-EGDMA sample with MS in the range 0.74–1 with water content reduced by compression to 20%.

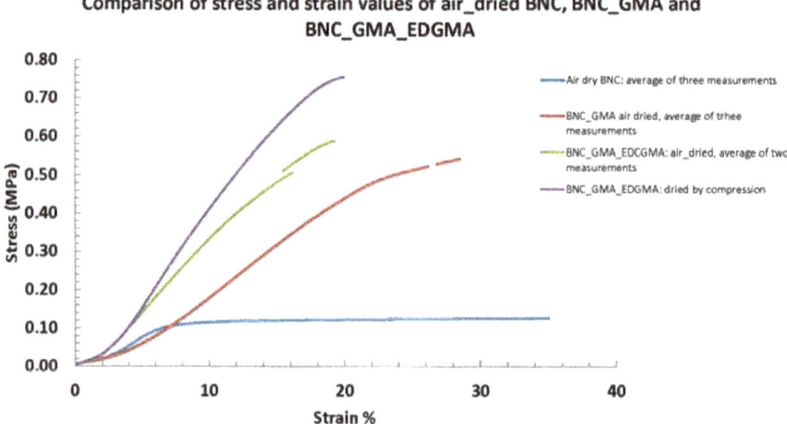

Figure 16. BNC (blue); BNC-GMA 0.64 MS (brown); BNC-GMA-EGDMA 0.7–1 MS (green and violet) stress and strain.

3.4. Adsorption and Release Antibacterial Activity Test

3.4.1. Adsorption and Release

The contact time between BNC and the drug solution was 24 h. Every hour the drug solution was UV analysed. No change occurred in the drug concentration. This observation allowed us to calculate the amount of drug adsorbed by the sample by diffusion from the drug solution into the hydrogel. The same drug concentration inside and outside the hydrogel was the equilibrium situation.

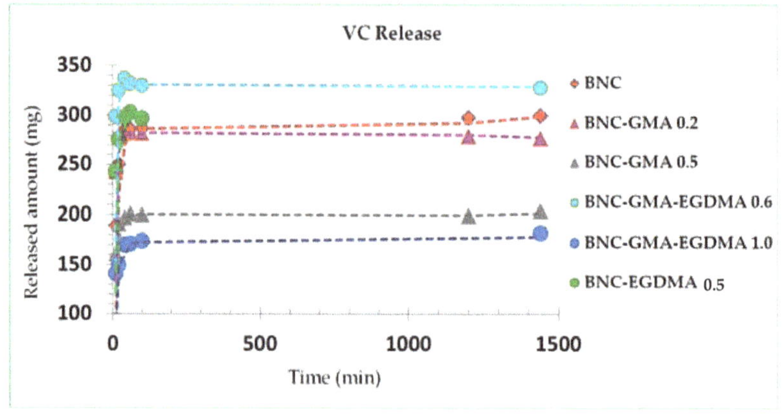

Figure 17. Kinetic of vancomycin (VC) release from BNC (red); BNC-GMA MS = 0.2 (violet) and MS = 0.5 (grey); BNC-EGDMA MS = 0.5 (green); BNC-GMA-EGDMA MS = 0.6 (light blue) and MS = 1.0 (blue).

In Figure 17 the amount of VC release from different BNC sample versus time measured by UV analysis at 280 nm is reported with the calibration curve in Figure 18. By elaboration on the experimental data and normalizing the value for 1 g of dry nanocellulose the equilibrium amount is 290 mg for VC, and 72 mg for CP.

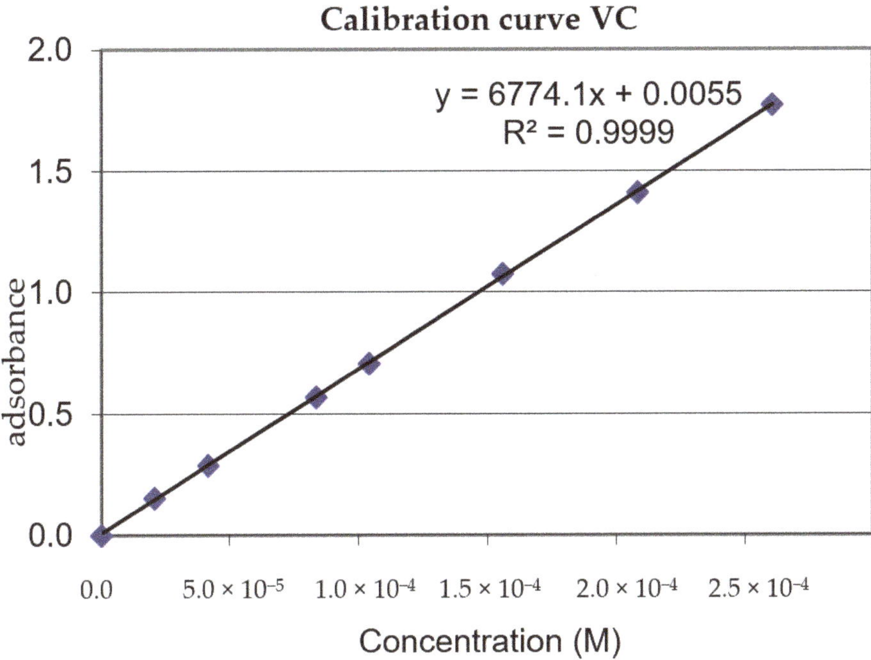

Figure 18. Calibration curve of VC at 272 nm, absorbance versus concentration.

In the graphic, the amount is reported for 1 g of dry nanocellulose. In 100 min all the samples reached a horizontal asymptote. BNC amount release is near 290 mg, thus supporting our interpretation of the absorption diffusive phenomenon. BNC-GMA 0.2 MS behaviour is quite similar to BNC. The cross-linked BNC-GMA-EGDMA with MS = 1 holds half of the adsorbed VC. The same behaviour is shown by BNC-GMA 0.5. By contrast, BNC-EGDMA 0.5 behaves as BNC. BNC-GMA-EGDMA with MS = 0.6 data could be attributed to an active adsorption, but it could be also considered in the range of the experimental error, suggesting that only highly cross-linked BNC-GMA-EGDMA really holds VC.

In Figure 19, the amount of CP release from different BNC sample versus time measured by UV analysis at 272 nm is reported, with the calibration curve in Figure 20. As with VC, in 100 min the three samples reached a horizontal asymptote. BNC amount release is near 80 mg, in good agreement with the theoretical value of 72. BNC-GMA with MS = 0.25 data could be attributed to an active CP adsorption. More evident is the behaviour of BNC-GMA-EGDMA with MS = 0.74, that holds 37% of the adsorbed CP.

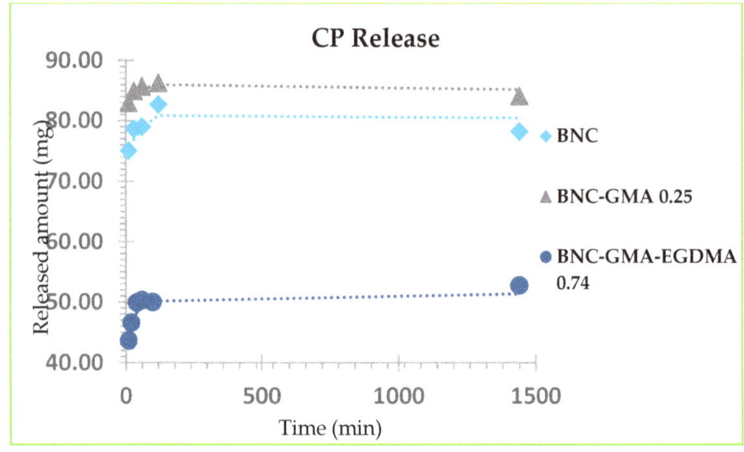

Figure 19. Kinetic of ciprofloxacin (CP) release from BNC (light blue); BNC-GMA MS = 0.25 (grey); BNC-EGDMA MS = 0.74 (blue).

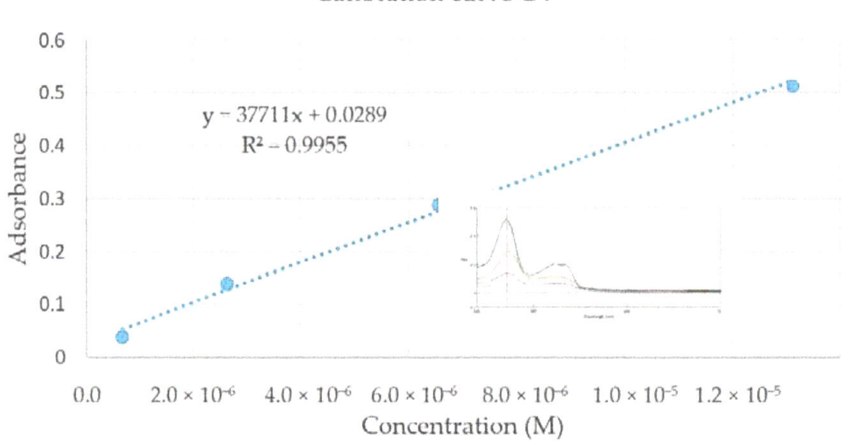

Figure 20. Calibration curve CP at 280 nm, absorbance versus concentration.

3.4.2. Antibacterial Activity Tests

BNC-GMA-EGDMA were used for antibacterial activity tests on the base of mechanical strength and release results. Four disks of BNC of 2.2 g were transformed in BNC-GMA-EGDMA with MS = 0.8–1, loaded with VC and CP according to the adsorption procedure and washed for 10 min in water. Four disks of BNC of 2.2 g were equally loaded with VC and CP according to the adsorption procedure and washed for 10 min in water, before the test. Every single disk was cut in four pieces and submitted to the in vitro tests described in paragraph 2. They did not undergo any degradation process. Their pictures are shown in Figure 21. The tests gave very positive results as all samples were not colonized by *Staphylococcus aureus* (*S. aureus*) and *Klebsiella pneumoniae* (*K. pneumonia*). The inhibition zone measurements are reported in Table 5. BNC zones of inhibition are larger than the corresponding BNC-GMA-EGDMA zone of inhibition. This behaviour appears in agreement with the release data and suggests the more long term antibacterial capability of BNC-GMA-EGDMA with respect to BNC.

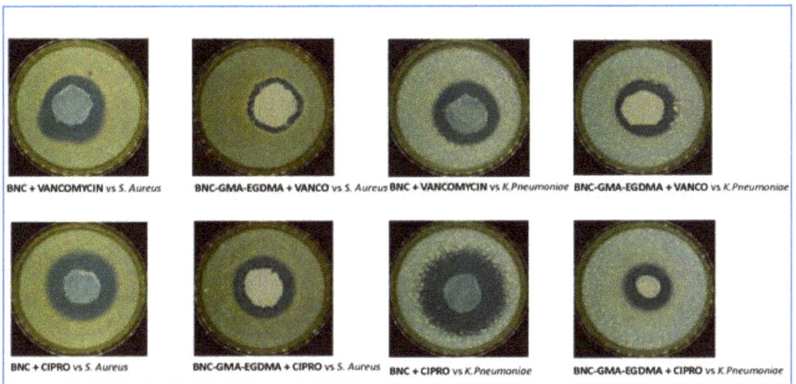

Figure 21. BNC and BNC-GMA-EGDMA MS = 0.8 antibacterial activity test. Left *S. aureus*, right *K. pneumoniae*. Upper VC, bottom CP.

Table 5. BNC and BNC-GMA-EGDMA antibacterial activity test, inhibition zone in mm in Figure 21.

Entry	Sample	VC	CP	*S. Aureus*	*K. Pneumoniae*	MS
1	BNC-GMA-EGDMA	x	-	4	-	0.8–1.0
2	BNC	x	-	8	-	-
3	BNC-GMA-EGDMA	-	x	8	-	0.8–1.0
4	BNC	-	x	10		-
5	BNC-GMA-EGDMA	x	-	-	5	0.8–1.0
6	BNC	x	-	-	7	-
7	BNC-GMA-EGDMA	-	x	-	11	0.8–1.0
8	BNC	-	x	-	15	-

4. Discussion

The study presented in this paper can be considered the first step towards a platform of new biomaterial based on BNC, built up with a method that was not yet investigated for nanocellulose. The results obtained with these studies start from the real possibility of applying to nanocellulose a process that was extensively studied for cellulose [16]. The peculiarity of this process is that it adds new properties to cellulose without modifying the original ones. The explanation for these results is the mechanism that involves two steps; the first is the generation by hydrogen abstraction of carbon centred macroradicals, sterically hindered and for this reason particularly stable. According to the polar effects involving in selective radical trapping process, these nucleophilic radicals are quickly trapped by double bonds substituted by electron withdrawing groups. Methacrylate groups are the ideal scavengers. The results of the scavenging process are the formation of a very stable C–C bond, and the maintenance of the glucose structure in the cellulose polymers that guaranties the maintenance of the cellulose properties. If many radicals are formed as the same time by a high reactive and unselective radical like OH· generated by a Fenton-type process, the scavenging step endows the cellulose surface of a network of methacrylate residues that is capable of capturing molecules like a sponge [16]. In collaboration with clinics of the Humanitas Hospital in Rozzano, Milano, Italy, we demonstrated that polysaccharide materials derivatised with GMA and loaded with antibiotics have topic antibacterial activity [17,18]. The clinics made in vitro and ex-vivo tests that stated the compatibility of GMA modified gauze with biological medium and with human skin. These results support the hypothesis that the same modification on nanocellulose leads to biocompatible materials. On the other hand the biocompatibility of BNC is well-established [6] and nanocellulose materials have been recently critically reviewed for drug delivery [27]. In particular they have shown considerable potential for developing a novel generation of controlled drug delivery for different routes of administration (oral, transdermal, etc.).

The transfer of this process to BNC was satisfying. As for the grafting on cellulose, the temperature is crucial in the first step in order to maintain the radical chain, as Fe(II) is used in catalytic amount and needs to be regenerated, and also in the second step to solubilize the methacrylate monomers in water. Entry 6 in Table 1 shows that the temperature cannot be reduced from 80 °C to 60 °C even with high reagents amounts. The Fenton-type process on BNC was successfully expanded to EGDMA, affording cross-linking that was not studied on cellulose. The cross-linking was finalised to enhance the resistance of BNC and to form on BNC a network with GMA of high loading capability and of long-term antibacterial property. The main drawback is the homogeneity when the starting BNC is of the order of magnitude of more than 2 g, due to the difficulties of efficient stirring. Future research directions will be the design of a suitable device to guarantee homogeneity, scale up, and economic evaluations of the process.

It is well-known that the properties of hydrogel state appear to be absolutely suitable to develop biomaterial in biomedical applications. The purpose of this study was to use both BNC and modified BNC in the hydrogel form. By contrast with many studies on BNC, we decided not to dry and rehydrate the samples, to reduce the step numbers to reach the final products. The percentage of nanocellulose in our modified BNCs increased in agreement with the MS, but our modified BNCs remained in the hydrogel state if the grafting and cross-linking reaction conditions were carefully controlled to not overcome MS = 1. At high MS, modified BNC became quite solid, due to water elimination that reached the 50%. Solid modified BNCs are not the achievement of this study, but they could be further developed as solid functionalised nanocelluloses of high purity.

SEM images and mechanical tests we presented cannot be compared with literature data that are mainly produced on dry samples. However mechanical tests of our modified BNC are very encouraging with respect to BNC as it is, in terms of resistance to the stress and strain. As expected, the cross-linking in particular has a positive effect on mechanical resistance, supporting the aim of our studies to make BNC cross-linked. SEM images are very interesting as it seems that GMA alone forms a dense thick network, while GMA-EGDMA network appears less dense. The density of the GMA network certainly helps the drug to be captured. GMA-EGDMA network is less dense, but due to the cross-linking is more rigid. This could induce a more stable electrostatic interaction between the GMA appendages and the drug. From a thermodynamic point of view we could argue that the loading enthalpy factors are more favourable for GMA entrapped in the cross-linking than for mobile GMA residues. At the same time, the BNC network was maintained. The observed pseudo-spherical structures could be explained by the formation of small pieces of cotton. It will be necessary to reconsider starting BNC preparation to avoid them.

It is very intriguing to comment on the crystallinity of the modified BNC that is maintained also at high MS equal to the BNC as it is. This behaviour confirms that the Fenton-type process we developed respects the morphology not only for cellulose, but also for nanocellulose. However, in our opinion the mechanism is not exactly the same. In the case of cellulose the functionalization occurs on the surface and in the bulk, while in the case of BNC it occurs on the fibre surface and does not modify at all the fibre itself.

To make the BNC bioactive we decided to use real antibiotics, with the idea that although antibiotics can develop resistance, they are the most potent tools we have to fight against infection. Hospital infections, surgical complications infections and wound infections are a tremendous problem for all of us and represent a relevant social cost, too.

BNC as it is appears suitable to host antibiotics like VC and CP, whose activities are well-established. VS is especially important in its use in the hospital.

Modified BNCs loaded with VC and CP can be developed as antibacterial materials. This conclusion is supported not only by adsorption release data collected in the laboratory, but also by the antibacterial activities tests entrusted to the expert Centrocot staff that use two very dangerous bacteria like S.A. and

K.P. BNC-GMA and BNC-GMA-EGDMA are not as available as BNC. However their properties justify the effort to prepare them. First of all, the mechanical properties are relevant. GMA appendages alone and associated with EGDMA cross-linking are capable of enhancing the BNC resistance. But likewise important is their interaction with the antibiotics. Efficient loading and holding depend on MS. BNC-GMA are efficient in the loading and holding of VC and CP, similarly to GMA modified cellulose. The only cross-linking with EGDMA does not seem efficient enough to hold drugs. The grafting and cross-linking are responsible for the most efficient holding capability. BNC-GMA-EGDMA behaviour suggests that there is a synergy of EGDMA and GMA to efficiently hold the drug. The synergy of EGDMA and GMA can be interpreted in the sense that GMA appendages create a network on a single cellulose chain and that EGDMA creates a network among different cellulose chains. Thus, BNC-GMA and BNC-GMA-EGDMA could open the way to a new kind of BNC, suitable to modulate the drug-delivery phenomenon in biomedical applications.

Author Contributions: Conceptualization, E.V., G.T.; writing—review and editing, E.V.; synthesis and analysis, A.B., C.B., S.P.; mechanic tests S.F.; SEM, A.S.; BNC technology, L.P., E.R.; NMR, G.T.

Funding: This research received no external funding.

Conflicts of Interest: The authors declare no conflict of interest.

References

1. Jorfi, M.; Foster, E.J. Recent advances in nanocellulose for biomedical applications. *J. Appl. Polym. Sci.* **2015**, *132*, 41719/1–41719/19. [CrossRef]
2. Gama, M.; Gatenholm, P.; Klemm, D. *Bacterial NanoCellulose: A Sophisticated Multifunctional Material*; CRC Press/Balkema: Leiden, The Netherlands, 2013; pp. 1–272.
3. Stumpf, T.R.; Yang, X.; Zhang, J.; Cao, X. In situ and ex situ modifications of bacterial cellulose for applications in tissue engineering. *Mater. Sci. Eng. C* **2018**, *82*, 372–383. [CrossRef] [PubMed]
4. Klemm, D.; Cranston, E.D.; Fischer, D.; Gama, M.; Kedzior, S.A.; Kralisch, D.; Kramer, F.; Kondo, T.; Lindstrom, T.; Nietzsche, S.; et al. Nanocellulose as a natural source for groundbreaking applications in materials science: Today's state. *Mater. Today* **2018**, *21*, 720–748. [CrossRef]
5. Sharma, C.; Bhardwaj, N.K. Bacterial nanocellulose: Present status, biomedical applications and future Perspectives. *Mater. Sci. Eng. C* **2019**, *104*, 109963. [CrossRef]
6. Osorio, M.; Canas, A.; Puerta, J.; Diaz, L.; Naranjo, T.; Ortiz, I.; Castro, C. Ex Vivo and In Vivo Biocompatibility Assessment (Blood and Tissue) of Three-Dimensional bacterial nanocellulose biomaterials for soft tissue implants. *Sci. Rep.* **2019**, *9*, 10553–10567. [CrossRef]
7. Bacakova, L.; Pajorova, J.; Bacakova, M.; Skogberg, A.; Kallio, P.; Kolarova, K.; Svorcik, V. Versatile Application of Nanocellulose: From Industry to Skin Tissue Engineering and Wound Healing. *Nanomaterials* **2019**, *9*, 164. [CrossRef]
8. Fu, L.; Zhang, Y.; Li, C.; Wu, Z.; Zhuo, Q.; Huang, X.; Qiu, G.; Zhou, P.; Yang, G. Skin tissue repair materials from bacterial cellulose by a multilayer fermentation method. *J. Mater. Chem.* **2012**, *22*, 12349–12357. [CrossRef]
9. Sheikhia, A.; Hayashia, J.; Eichenbauma, J.; Gutina, M.; Kuntjoroa, N.; Khorsandia, D.; Ali Khademhosseinia, A. Recent advances in nanoengineering cellulose for cargo delivery. *J. Control. Release* **2019**, *294*, 53–76. [CrossRef]
10. Liyaskina, E.; Revin, V.; Paramonova, E.; Nazarkina, M.; Pestov, N.; Revina, N.; Kolesnikova, S. Nanomaterials from bacterial cellulose for antimicrobial wound dressing. *J. Phys. Conf. Ser.* **2017**, *784*, 012034. [CrossRef]
11. Shao, W.; Liua, H.; Wang, S.; Wu, J.; Huang, M.; Min, H.; Liu, X. Controlled release and antibacterial activity of tetracycline Hydrochloride-Loaded bacterial cellulose composite membranes. *Carbohydr. Polym.* **2016**, *145*, 114–120. [CrossRef]
12. Moritz, S.; Wiegand, C.; Wesarg, F.; Hessler, N.; Müller, F.A.; Kralisch, D.; Hipler, U.-C.; Fischer, D. Active wound dressings based on bacterial nanocellulose as drug delivery system for octenidine. *Int. J. Pharm.* **2014**, *471*, 45–55. [CrossRef]

13. Trovatti, E.; Silva, N.H.C.S.; Duarte, I.F.; Rosado, C.F.; Almeida, I.F.; Costa, P.; Freire, C.S.R.; Silvestre, A.J.D.; Neto, C.P. Biocellulose membranes as supports for dermal release of lidocaine. *Biomacromolecules* **2011**, *12*, 4162–4168. [CrossRef] [PubMed]

14. Sannino, A.; Demitri, C.; Madaghiele, M. Biodegradable Cellulose-based Hydrogels: Design and Applications. *Materials* **2009**, *2*, 353–373. [CrossRef]

15. Torres, F.G.; Arroyo, J.J.; Troncoso, O.P. Bacterial cellulose nanocomposites: An all-nano type of material. *Mater. Sci. Eng. C* **2019**, *98*, 1277–1293. [CrossRef]

16. Vismara, E.; Melone, L.; Torri, G. *Surface Functionalizationed Cotton with Glycidyl Methacrylate: Physico-Chemical Aspects and Multitasking Applications*; Giuliano, B., Vinci, E.J., Eds.; Cotton: Cultivation, Varieties and Uses; Nova Science Publishers: New York, NY, USA, 2012; Chapter 3; pp. 125–147.

17. Vismara, E.; Torri, G.; Valerio, A.; Graziani, G.; Montanelli, A.; Melone, L. Nanostructured cellulose materials: Adsorption of antibiotics onto cellulose fibers functionalized with glycidylmethacrylate for the manufacturing of antibacterial fabrics. In Proceedings of the 2012 NSTI Nanotechnology Conference and Trade Show, Santa Clara, CA, USA, 18–21 June 2012; Volume 3, Chapter 3. pp. 174–177.

18. Graziani, G.; Montanelli, A.; Melone, L.; Vismara, E.; Torri, G. Derivatised Polysaccharide Material for the Topic Antibacterial Activity. European Patent 2182931 B1, 20 April 2016.

19. Rijith Sreenivasana, R.; Suma Mahesha, S.; Sumib, V.S. Synthesis and application of polymer-grafted nanocellulose/graphene oxide nano composite for the selective recovery of radionuclides from aqueous media. *Sep. Sci. Technol.* **2019**, *54*, 1453–1468. [CrossRef]

20. Hiratani, H.; Mizutani, Y.; Alvarez-Lorenzo, C. Controlling Drug Release from Imprinted Hydrogels by Modifying the Characteristics of the Imprinted Cavities. *Macromol. Biosci.* **2005**, *5*, 728–733. [CrossRef]

21. Pacheco, H.; Vedantham, K.; Aniket; Young, A.; Marriott, I.; El-Ghannam, A. Tissue engineering scaffold for sequential release of vancomycin and rhBMP2 to treat bone infections. *J. Biomed. Mater. Res. Part A* **2014**, *102A*, 4213–4223. [CrossRef]

22. Yu, W.; Sun, T.W.; Qi, C.; Ding, Z.; Zhao, H.; Chen, F.; Chen, D.; Zhu, Y.J.; Shi, Z.; He, Y. Strontium-doped amorphous calcium phosphate porous microspheres synthesized through a microwave-hydrothermal method using fructose 1,6-bisphosphate as an organic phosphorus source: Application in drug delivery and enhanced bone regeneration. *ACS Appl. Mater. Interfaces* **2017**, *9*, 3306–3317. [CrossRef]

23. Verma, N.K.; Ramakrishna, S.; Lakshminarayanan, R. Bio-inspired cross-linking and matrix-drug interactions for advanced wound dressings with long-term antimicrobial activity. *Biomaterials* **2017**, *138*, 153–168.

24. Khattab, M.M.; Dahman, Y. Functionalized bacterial cellulose nanowhiskers as long-lasting drug nanocarrier for antibiotics and anticancer drugs. *Can. J. Chem. Eng.* **2019**, 1–14. [CrossRef]

25. Vismara, E.; Melone, L.; Gastaldi, G.; Cosentino, C.; Torri, G. Surface functionalization of cotton cellulose with glycidyl methacrylate and its application for the adsorption of aromatic pollutants from wastewaters. *J. Hazard. Mater.* **2009**, *170*, 798–808. [CrossRef] [PubMed]

26. Schneider, C.A.; Rasband, W.S.; Eliceiri, K.W. NIH Image to ImageJ: 25 years of image analysis. *Nat. Methods* **2012**, *9*, 671–675. [CrossRef] [PubMed]

27. Salimi, S.; Sotudeh-Gharebagh, R.; Zarghami, R.; Chan, S.Y.; Yuen, K.H. Production of nanocellulose and its applications in drug delivery: A critical review. *ACS Sustain. Chem. Eng.* **2019**, *7*, 15800–15827. [CrossRef]

 *nanomaterials*

Article

A Highly Efficient and Durable Fluorescent Paper Produced from Bacterial Cellulose/Eu Complex and Cellulosic Fibers

Mingquan Zhang [1], Xiao Wu [1], Zhenhua Hu [1], Zhouyang Xiang [1,*], Tao Song [1] and Fachuang Lu [1,2,*]

1   State Key Laboratory of Pulp and Paper Engineering, South China University of Technology, Guangzhou 510640, China; fezhangmingquan@mail.scut.edu.cn (M.Z.); msxiaowu@mail.scut.edu.cn (X.W.); zhenhuahu1991@163.com (Z.H.); songt@scut.edu.cn (T.S.)
2   Guangdong Engineering Research Center for Green Fine Chemicals, Guangzhou 510640, China
*   Correspondence: fezyxiang@scut.edu.cn (Z.X.); fefclv@scut.edu.cn (F.L.)

Received: 12 August 2019; Accepted: 12 September 2019; Published: 15 September 2019

Abstract: The general method of producing fluorescent paper by coating fluorescent substances onto paper base faces the problems of low efficiency and poor durability. Bacterial cellulose (BC) with its nanoporous structure can be used to stabilize fluorescent particles. In this study, we used a novel method to produce fluorescent paper by first making Eu/BC complex and then processing the complex and cellulosic fibers into composite paper sheets. For this composting method, BC can form very stable BC/Eu complex due to its nanoporous structure, while the plant-based cellulosic fibers reduce the cost and provide stiffness to the materials. The fluorescent paper demonstrated a great fluorescent property and efficiency. The ultraviolet absorbance or the fluorescent intensity of the Eu-BC fluorescent paper increased with the increase of Eu-BC content but remained little changed after Eu-BC content was higher than 5%. After folding 200 times, the fluorescence intensity of fluorescent paper decreased by only 0.7%, which suggested that the Eu-BC fluorescent paper has great stability and durability.

Keywords: bacterial cellulose; Eu ion; complex; cellulosic fiber; fluorescent paper; durability

---

1. Introduction

Europium (Eu) element has one of the best photoluminescent properties among rare earth elements. Due to the 4f→4f transition, Eu^{3+} ions can convert ultraviolet light into visible light and have been widely used in fluoroimmunoassay [1,2]. The antenna effect induced by the chelation between Eu^{3+} and negatively charged ligand results in an even stronger photoluminescent performance for ligand/Eu^{3+} complexes. The complexes of various ligands, e.g., β-diketone, chelated with Eu^{3+} can be used as photoreceptors, superconductors, magnetic materials, catalysts, fluorescent probes, and light-emitting diodes [3–8].

Complexes of polymer ligands and rare earth elements are considered as high-efficiency light-emitting materials because of their strong photoluminescent ability, easy color mixing, low temperature sensitivity, and high thermal stability [9]. In the complex of cholesterol-g-poly (N-isopropylacrylamide) (PNIPAM) and Eu^{3+}, the oxygen-nitrogen coordination between amide group and Eu^{3+} provides a strong fluorescent characteristic for Eu^{3+} [10]. Those ligands with functional groups containing strong electron-donating groups, e.g., hydroxyls, carboxyls, amides, and nitriles, especially have strong chelation effects with Eu^{3+} ions or other rare earth elements. Carboxymethyl cellulose (CMC) or cholesterol-g-poly(N-isopropylacrylamide) were used to complex with Eu or Tb, and they showed strong photofluorescent properties [11–14]. Hybridized nanofibers were produced between polyacrylonitrile and Eu^{3+}, which had one-dimensional nanostructure and high specific surface area, showing excellent fluorescent properties [1]. However, most of these photofluorescent complexes or materials are still facing the problems of stability and durability, i.e., the fluorescence intensity decreases after severe use, due to the leaching or aggregation of fluorescent elements or particles.

When the polymer ligands reduce to nano- or submicron-scale, they provide the materials with special characteristics in mechanical, optical, thermal, or surface properties [15]. Nanocellulose, with its sustainable nature and nanoscale characteristics, inspires its application in functional materials. So far, only a few studies have used nanocellulose, e.g., cellulose nanocrystal (CNC), cellulose nanofibril (CNF), and bacterial cellulose (BC), as supports or ligands for fluorescent substances. Fluorescein-5′-isothiocyanate (FITC) was used to label CNCs for fluorescence bioassay and bio-imaging applications [16]. CNCs grafted by poly (N-isopropylacryalamide)(PNIPAAM) brushes had thermo-enhanced fluorescence and can be used in biomedical applications [17]. The silver nanocluster was loaded on natural CNFs, giving the material fluorescence properties and antimicrobial properties [18]. A fluorescent film was prepared by chemical modification of CNF film surface with boronate-terminated conjugated polymer chains and used to detect nitro-aromatic vapor [19].

Bacterial cellulose (BC) is a type of typical extracellular cellulose produced by acetobacter bacteria. BC was composed of ultrafine and interlaced nanofibers with a width of less than 100 nm to form a network structure. This structural feature provides BC with a large specific surface area and a nanoporous structure [20,21]. This nanoporous structure makes BC an ideal support for functional nanoparticles that can be further processed into functional materials. BC membranes were used to support TiO_2 nanoparticles and blended with rare earth elements, e.g., lanthanum (La) and cerium (Ce), leading to a high photocatalytic efficiency [22]. BC and CdSe were combined to prepare composite films which have photoluminescence properties [23]. BC was used to support nitrogen-doped carbon nanofibers; this composite showed electrode material properties and was used to prepare a flexible all-solid-state high-power supercapacitor [24]. Pt was supported by BC to prepare a composite material having high catalytic efficiency for methanol fuel battery [25]. There are very few studies on the preparation of photofluorescent materials by using BC to support and form complexes with rare earth elements. The possible reason is the high cost of BC and its soft nature upon film forming, which is not able to provide a good stiffness for the materials.

Plant-based cellulosic fibers are well known to be the raw material for paper making and they can also be used as substrates or supporting materials for functional ligands to produce functional paper-based materials. Fluorescent paper has also been produced. Coating or printing method is a common way to composite the fluorescent materials with paper sheets. Fluorescent DNA-based oligodeoxyfluoroside dyes were printed on paper, which can sense food spoilage and ripening in the vapor phase [26]. 8-hydroxyquinoline aluminum (Alq3)-based bluish green fluorescent composite nanospheres were printed on paper as sensors for nitroaromatic explosive detection [27]. A ratiometric fluorescent probe-based paper sensor based on paper printed with gold nanoclusters stabilized by bovine serum albumin and fluorescent graphene oxide can be used for the determination of serum blood sugar [28]. Fluorescent inks such as heterorotaxane and the mixture of triethanolamine and UV absorbent UV-7282 were also used to print on paper for fluorescence [29,30]. Fluorescein isothiocyanate (FITC) and thioflavin were absorbed by nanocellulose and that was coated on the surface of paper to prepare fluorescent paper [31]. Most of the current methods produce fluorescent paper by attaching fluorescent materials to the surface of paper without altering the paper fiber structure. However, common cellulosic fibers do not have nanoporous structure and thus cannot well stabilize the nanosized fluorescent particles, leading to lack of stability, recycling efficiency, and durability.

BC can be tightly bound to cellulosic fibers due to their abundant surface hydroxyl groups [20,21], whose features can be used to prepare functional paper-based materials if BC is properly functionalized. In cellulosic fiber/BC composite functional materials, cellulosic fibers can reduce the cost of pure BC materials and provide high stiffness, while BC can still provide the materials with high toughness, high tearing, good stability, and durability. Catalytic paper sheets were produced by loading Pt or Au onto BC surface and then compositing with cellulosic fibers, which showed excellent catalytic efficiency, stability, and reusability [32,33]. In this study, we prepared fluorescent paper by making

BC/Eu complex first and then processing the complex and cellulosic fibers into paper sheets. This method incorporates the fluorescent complexes as part of the cellulosic fiber matrix and thus may improve the stability and durability of fluorescent paper.

2. Materials and Methods

2.1. Materials

Gluconacetobacter xylinus ATCC23767 was obtained from Nanjing High Tech University Biological Technology Research Institute Co., Ltd. (Nanjing, China) and used to produce the bacterial cellulose (BC) pellicles according to the static fermentation method [20,21]. The degree of polymerization (DP) and Segal crystallinity index (CI) of BC was predetermined as 1081 and 96.0%, respectively [20,21].

Europium trioxide (Eu_2O_3), hydrochloric acid, and sodium hydroxide were purchased from Shanghai Macklin Biochemical Co., Ltd. (Shanghai, China). All other chemicals are analytical grade chemicals.

2.2. Preparation of EuCl$_3$·6H$_2$O

Eu_2O_3 (0.727 g) was completely dissolved in 6.0 mol/L HCl solution. The mixture was placed in boiling water bath for 30 min to evaporate the unreacted HCl. The solution was then diluted with deionized water to volume 100 mL to adjust the concentration to 0.0416 mol/L.

2.3. Preparation of Eu-BC

Approximately 20 g of wet BC pellicles (moisture content = 98.5%) was disintegrated by a lab homogenizer (SKG 1246, Foshan, China) for 20 s in 130 mL of water. The water was then filtered to obtain the disintegrated BC. The pH of the $EuCl_3$·$6H_2O$ solution was adjusted to 5–6 with 1 M NaOH. The disintegrated BC was then added slowly to the $EuCl_3$·$6H_2O$ solution at 45 °C under a magnetic stirring (400 rpm). The pH of mixture was adjusted to about 7 with 1 M NaOH. Finally, the mixture was refluxed at 70 °C for 1 h 30 min to prepare Eu-BC.

2.4. Manufacture of Fluorescent Paper

A vacuum filtering method was used to prepare the fluorescent paper because at high EuBC addition it would take a very long time at the water drainage step for the standard TAPPI method. In detail, bleached sugarcane bagasse pulps (BSBP) and Eu-BC were composited with different ratios, i.e., 1%, 2%, 5%, 10%, and 20% (the proportion of Eu-BC based on total paper dry weight). The mixture was homogenizer for 20 s and the mixture was filtered by a sand core filter device. After water was drained, the paper sheet was formed. Finally, the paper was dried in an oven at 9 °C for 50 min.

2.5. Characterization of Eu-BC and Fluorescent Paper

Nitrogen adsorption–desorption isotherm measurements were performed on an ASAP 2460 (Micromeritics, Norcross, GA, USA) volumetric adsorption analyzer at 77 K. The Brunauer–Emmett–Teller (BET) method was utilized to calculate the specific surface area of each sample. All samples were degassed at 120 °C at 0.05 atm for 6 h before the measurement began.

Fluorescent paper sheets or Eu-BC were taken under an AGL-9406 UV lamp for excitation at 256 nm. The ultraviolet light absorption of the samples was measured by solid-state ultraviolet spectrophotometer (UV-2450, SHIMADZU, Kyoto, Japan). The fluorescence spectra of the samples were recorded by a steady-state transient fluorescence spectrometer (FluoroMax-4, HORIBA Jobin Yvon, Longjumeau, France).

The Eu content of the paper was measured by an atomic absorption spectrometer (Z-2000, Hitachi, Tokyo, Japan). The surface morphologies of unmodified BC, Eu-BC, Eu-BC fluorescent papers were observed by a scanning electron microscope (SEM) (Merlin, Zeiss, Munich, Germany); X-ray photoelectron spectroscopy (XPS) was conducted on the fluorescent paper by a X-ray photoelectron spectrometer (Kratos Axis Ulra DLD, Kratos Analytical, Manchester, UK) to analyze the valence and binding energy of the Eu ions.

3. Results and Discussion

3.1. Characterization of Eu-BC and Fluorescent Paper

Bacterial cellulose (BC) is a natural cellulosic material with nanoporous structure, which provides the advantages to stably support functional nanoparticles. The BC used in this study was prepared by static fermentation method instead of agitated method because BC prepared by static method provides paper with better mechanical properties [21]. The BC had a specific surface area of 41.69 m^2/g and a pore volume of 0.1598 cm^3/g (Table 1). As can be seen from the SEM image (Figure 1a), BC is composed of ultrafine and reticular nanofibrils demonstrating a large specific surface area. Eu-BC complex was prepared by adsorbing Eu^{3+} onto the nanoporous structure of BC. The total Eu content of Eu-BC was 32.3 wt% (Table 1), which suggested that the Eu ions were successfully adsorbed onto BC. The XPS spectrum of Eu-BC showed that the binding energies of the two peaks were close to 1126 and 1155 eV, which belonged to the Eu^{3+} valence state and suggested that all the Eu in BC was still in Eu^{3+} forms (Figure 2a). The XPS spectrum indicated that Eu^{3+} ions were successfully coordinated with BC to form a Eu-BC complex. Specific surface area and pore volume of Eu-BC decreased greatly to 6.84 m^2/g and 0.0214 cm^3/g, indicating Eu^{3+} formed uniform complex with BC and occupied most of pores on BC surface. As demonstrated by SEM image (Figure 1b), Eu-BC complex was evenly distributed or dispersed among the BC matrix. Eu-BC complex was evenly dispersed because of the nanoporous structure of BC, and it also indicated the formation of nanosized complex particles.

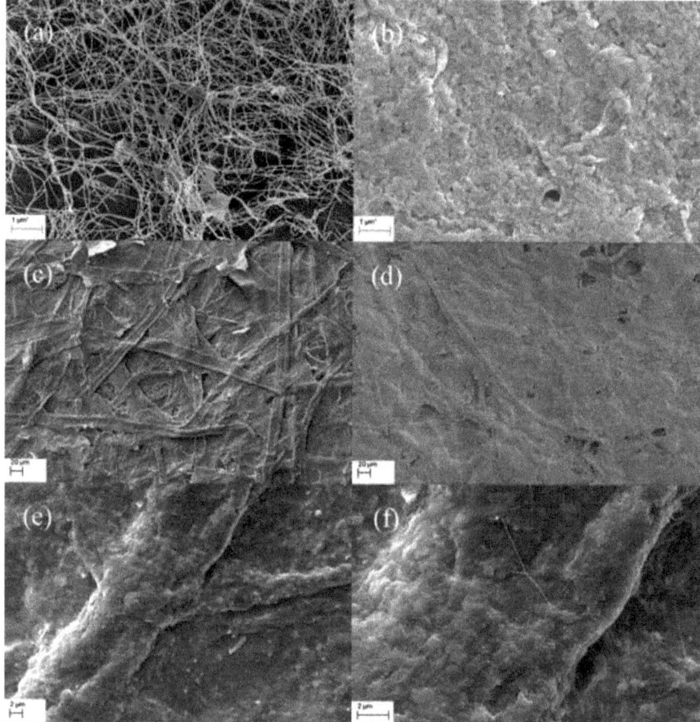

Figure 1. SEM images of (**a**) unmodified BC, (**b**) Eu-BC, (**c**) paper sheet made from sugarcane bagasse pulp, (**d**) Eu-BC fluorescent paper, and the cellulosic fibers on Eu-BC fluorescent paper at (**e**) 2K× magnification and at (**f**) 5K× magnification.

The Bleached sugarcane bagasse pulp (BSBP) was composited with Eu-BC to produce fluorescent papers with different Eu-BC contents. The Eu content in Eu-BC fluorescent paper as determined by atomic absorption spectrometry was proportional to the amount of Eu-BC added to the production of the fluorescent paper, demonstrating a uniform adsorption of Eu ions onto BC (Table 1). Most of the Eu elements in the fluorescent paper were still in the valence state of 3+ (Figure 2b), facilitating the photofluorescence of the fluorescent paper. The fluorescent paper showed red fluorescence under ultraviolet light irradiation at 256 nm (Figure 3), indicating that the energy of ultraviolet light absorbed is effectively transferred to Eu^{3+} through the "antenna effect", making the fluorescent paper glow red [11–14]. It can be seen from the SEM images that almost all the bagasse fibers were uniformly covered by Eu-BC (Figure 1c–e). The reticular structure of BC can also be identified on fiber surface (Figure 1f).

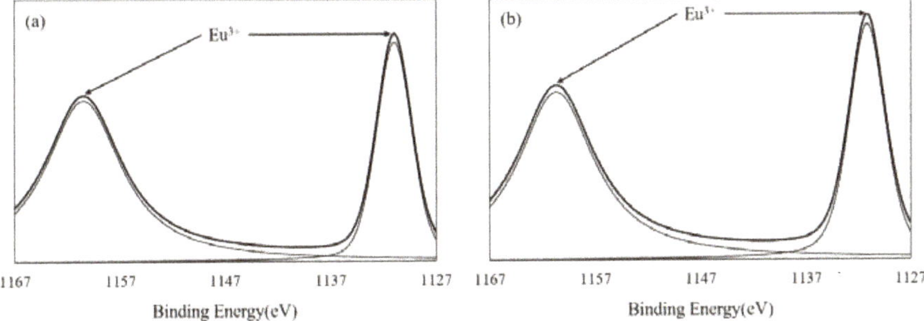

Figure 2. XPS spectra of (**a**) Eu-BC and (**b**) 20% Eu-BC fluorescent paper.

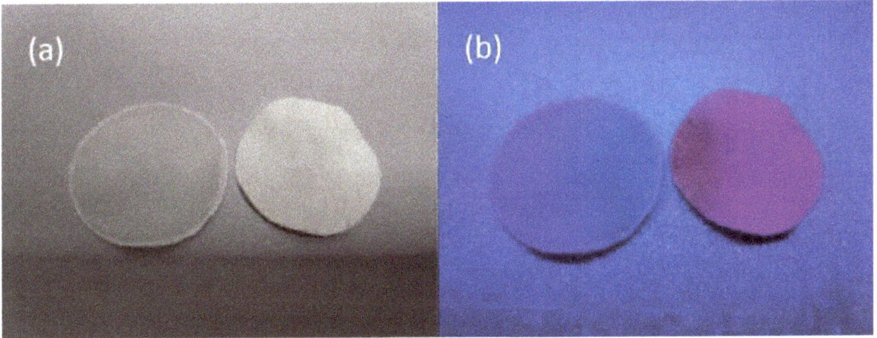

Figure 3. Photographs of 20% BC paper (without Eu) and 20% Eu-BC fluorescent paper (**a**) under visible light and (**b**) under ultraviolet light.

Table 1. Eu contents and BET parameters of BC, Eu-BC, and fluorescent paper sheets with different Eu-BC contents.

Sample	Eu wt%	Specific Surface Area (m²/g)	Pore Volumn (cm³/g)	Average Pore Radius (nm)
BC	0	41.69	0.1598	15.81
Eu-BC	32.30	6.84	0.0214	15.62
1% Eu-BC fluorescent paper	0.32	6.81	0.0991	79.84
2% Eu-BC fluorescent paper	0.88	-	-	-
5% Eu-BC fluorescent paper	1.53	3.17	0.0024	5.81
10% Eu-BC fluorescent paper	2.44	-	-	-
20% Eu-BC fluorescent paper	7.74	12.12	0.0430	14.74

3.2. Photofluorescent Properties of Eu-BC and Eu-BC Paper

As can be seen from Figure 4, there were two peaks in the Eu-BC excitation spectrum, 305 nm and 312 nm (Figure 4a). The 305 nm wavelength was selected to excite fluorescent papers of different Eu-BC contents to obtain a fluorescence emission spectrum. With the excitation at 305 nm wavelength, the fluorescent paper produced radiation emission at 605 nm and 618 nm (Figure 4b), which should correspond to Eu^{3+} magnetic dipole transition of $^5D_0 \rightarrow {}^7F_1$ and electric dipole transition of $^5D_0 \rightarrow {}^7F_2$ [1]. The peaks of Figure 4b suggested that the emission intensity of fluorescent paper below 5% Eu-BC content increases with the increases of Eu-BC content. The emission intensity of fluorescent paper conducted no more increases when the EU-BC content in fluorescent paper was greater than 5%.

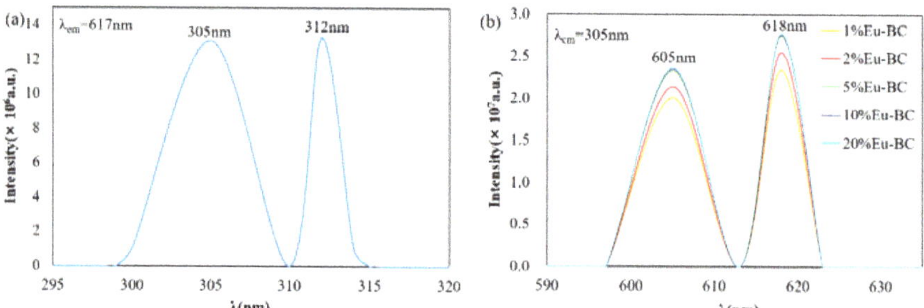

Figure 4. (**a**) Fluorescence excitation spectra of Eu-BC and (**b**) fluorescence emission spectra of fluorescent paper sheets with different Eu-BC contents.

By evaluating the solid-state ultraviolet absorption spectrum (Figure 5a), the fluorescent paper had two absorbance peaks at the wavelength of ~230 and ~280 nm. The absorbance of 1%, 2%, and 5% Eu-BC fluorescent paper successively increased with the increase of Eu-BC content. However, the absorbance of 5%, 10%, and 20% Eu-BC fluorescent paper remained almost unchanged. This result was consistent with the fluorescence excitation spectrum, showing that after Eu-BC content exceeds 5% the fluorescent property would not change for the fluorescent paper.

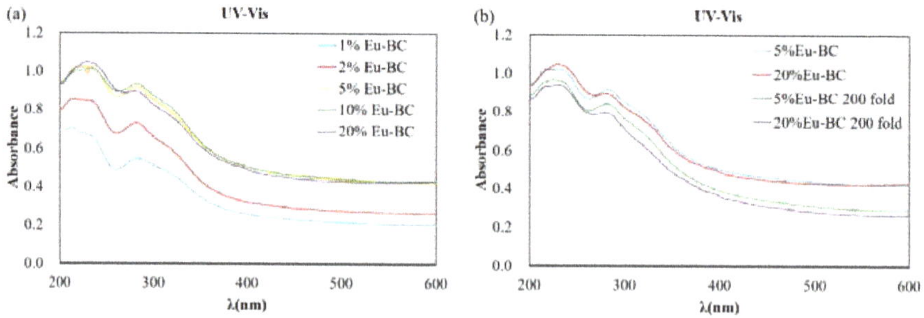

Figure 5. (**a**) UV-Vis absorbance spectra of fluorescent paper sheets with different Eu-BC contents and (**b**) changes in UV-Vis spectra after fluorescent paper was folded 200 times.

The structure of 5% Eu-BC and 20% Eu-BC fluorescent paper were shown in Figure 6. For the 5% Eu-BC paper, the reticular structure of BC can still be seen, but for the 20% Eu-BC paper, the paper was all covered with Eu-BC complex packed full with Eu elements, in which some Eu-BC complexes were observed to be stacked with each other. Many Eu-BC complexes or Eu particles were squeezed or buried inside the middle of the composite, which made them hard to be stimulated by ultraviolet light. This statement can be proved by the surface area, pore volume, and pore size data of EuBC paper (Table 1). For 1% EuBC paper, the cellulosic fibers were not entirely covered with EuBC, so it had relatively high pore volume and pore size; for 5% EuBC paper, the cellulosic fibers were largely covered with EuBC, so its pore volume and pore size decreased; for 20% EuBC paper, many EuBC stacked together, building some now pores between layers leading to the increase in pore volume and pore size (Table 1). Similar results were also found in the studies of poly(acrylonitrile)/Eu^{3+} complex that high Eu^{3+} does not necessary lead to improved emission intensity due to the stacking of emission centers [1]. This explains the fluorescence intensity or ultraviolet absorbance remaining constant when Eu-BC is over 5%. It also indicated that 5% is the most economical amount for Eu-BC to composite with cellulosic fibers.

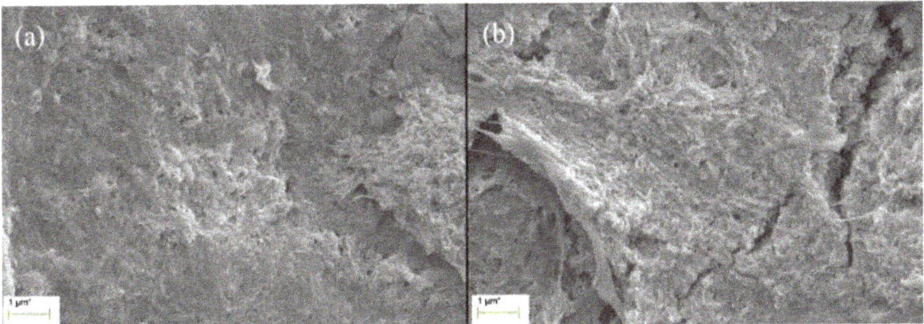

Figure 6. SEM images of (**a**) 5% Eu-BC fluorescent paper and (**b**) 20% Eu-BC fluorescent paper at 10K× magnification.

3.3. Stability and Durability of Fluorescent Paper

To investigate the stability and durability of fluorescent paper, 5% Eu-BC fluorescent paper and 20% Eu-BC fluorescent paper were repeatedly folded 200 times. From the ultraviolet absorbance spectrum in Figure 5b, it can be found that the absorbance peaks decreased only slightly after folding 200 times. The absorbance of 5% Eu-BC fluorescent paper decreased by only 0.05 after 200 folds, while the 20% Eu-BC paper decreased by nearly 0.1. The fluorescence intensity at 618 nm emission changes after the folding of the fluorescent paper were shown in Figure 7. The intensity of the fluorescent paper only decreased by ~0.7% after 200 folds. This result suggested that the fluorescent paper has great stability and durability, and 5% Eu-BC fluorescent paper is even better than the 20% Eu-BC fluorescent paper. It suggested that the nanoporous and reticular structure of BC can help to form very stable complex with Eu ions, by preventing the Eu particles from leaching or aggregating.

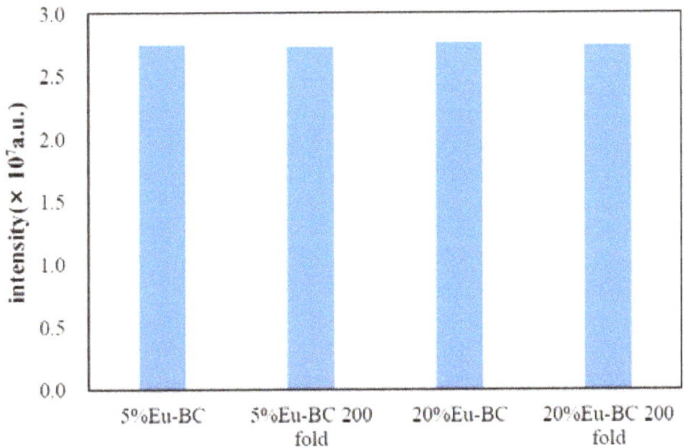

Figure 7. Changes of fluorescence emission spectra at 618 nm for Eu-BC fluorescent paper after 200 foldings.

3.4. Comparison of Eu-BC Fluorescent Paper with Other Composite Fluorescent Materials

We mainly compared the fluorescence intensity of different substrates complexing with Eu^{3+} ions, such as CMC [12], polydimethylsiloxane (PDMS) [34], β-diketone,4-imidazol-4,4,4-trifluorobutane-1,3-dione(HIDTFBD) [3], 2-(4′4′4′-trifluoro-1′3′-dioxobutyl)-carbazole (2-TFDBC) [35], and 2,7-bis (4′4′4′-trifluoro-1′3′-dioxobutyl)-carbazole (2,7-BTFDBC) [35]. The highest fluorescent emission intensity of the Eu-BC fluorescent paper in this study can reach 2.77×10^7 a.u., showing higher value than many of the Eu-ligand complex fluorescent materials; the Eu-BC fluorescent paper also had a lowest Eu mass ratio among all showing its high efficiency (Table 2). This may be due to the high surface area of BC providing large numbers of chelating centers for Eu^{3+} and the nanoporous structure of BC providing uniform distribution of Eu elements. In addition, our Eu-BC fluorescent paper maintained good fluorescence performance after 200 folds. These comparisons with similar studies demonstrate the high efficiency and great durability for our Eu-BC fluorescent paper.

Table 2. Comparison of composites with different substrate loadings of Eu^{3+} ions.

Base	Reaction Temperature (°C)	Reaction Time (h)	Eu Mass Percentage	Fluorescent Emission Intensity (a.u.)	Reference
BC/Paper	70	1.5	1.5%	2.77×10^7	this work
CMC	70	0.25	4.4%	1.05×10^7	[12]
HIDTFBD	60	6	18.0%	4.21×10^6	[3]
PDMS	60	3	2%	1.42×10^4	[34]
2-TFDBC	60	6	16.7%	6.90×10^6	[35]
2,7-BTFDBC	60	6	24.4%	6.20×10^6	[35]

4. Conclusions

Fluorescent paper was prepared by compositing the Eu-BC complex with sugarcane bagasse pulps (BSBP). The fluorescent paper demonstrated a great fluorescent property and efficiency, i.e., having very low Eu mass content but high fluorescent intensity, compared to many of its counterpart fluorescent materials based on Eu/ligand complex. The ultraviolet absorbance or the fluorescent intensity of the Eu-BC fluorescent paper increased with the increase of Eu-BC content but remained little changed after Eu-BC content was higher than 5%, which was due to excessive Eu-BC complexes stacking with each other without effective irritation by ultraviolet light. After 200 times of folding, the fluorescence intensity of fluorescent paper decreased by only 0.7%, which suggested that the Eu-BC fluorescent paper has very good stability and durability. The high durability of the fluorescent paper can be attributed to the nanoporous or reticular structure of BC stabilizing the Eu particles.

Author Contributions: Conceptualization, Z.X.; methodology, Z.X.; investigation, M.Z. and X.W.; resources, T.S.; writing—original draft preparation, M.Z.; writing—review and editing, Z.X. and Z.H.; supervision, F.L.; project administration, Z.X.; funding acquisition, Z.X. and F.L.

Funding: This work was supported by Guangzhou Science and Technology Program (Key Scientific Research Project 201707020011), the National Natural Science Foundation of China (31600470), Guangzhou Science and Technology Program (General Scientific Research Project 201707010053), and Guangdong Province Science Foundation for Cultivating National Engineering Research Center for Efficient Utilization of Cellulosic fibers (2017B090903003).

Acknowledgments: The authors would like to thank Nanjing High Tech University Biological Technology Research Institute Co., Ltd. for providing the bacterial cellulose.

Conflicts of Interest: The authors declare no conflict of interest.

References

1. Tang, S.; Shao, C.; Liu, Y.; Mu, R. Electrospun nanofibers of poly(acrylonitrile)/Eu3+ and their photoluminescence properties. *J. Phys. Chem. Solid.* **2010**, *71*, 273–278. [CrossRef]
2. Sabbatini, N.; Guardigli, M.; Lehn, J.M. Luminescent lanthanide complexes as photochemical supramolecular devices. *Coord. Chem. Rev.* **1993**, *123*, 201–228. [CrossRef]

3. Wang, H.; He, P.; Yan, H.; Shi, J.; Gong, M. A novel europium(III)–imidazol–diketonate–phenanthroline complex as a red phosphor applied in LED. *Inorg. Chem. Commun.* **2011**, *14*, 1183–1185. [CrossRef]

4. Reyes, R.; Cremona, M.; Teotonio, E.E.S.; Brito, H.F.; Malta, O.L. Voltage color tunable OLED with (Sm,Eu)-β-diketonate complex blend. *Chem. Phys. Lett.* **2004**, *396*, 54–58. [CrossRef]

5. Teotonio, E.; Brito, H.F.; da Cunha Felinto, M.C.F.; Thompson, L.C.; Young, V.G.; Malta, O.L. Preparation, crystal structure and optical spectroscopy of the rare earth complexes (RE3$^+$ = Sm, Eu, Gd and Tb) with 2-thiopheneacetate anion. *J. Mol. Struct.* **2005**, *751*, 85–94. [CrossRef]

6. Reineke, T.M.; Eddaoudi, M.; Fehr, M.; Kelly, D.; Yaghi, O.M. From condensed lanthanide coordination solids to microporous frameworks having accessible metal sites. *J. Am. Chem. Soc.* **1999**, *121*, 1651–1657. [CrossRef]

7. Ma, L.; Evans, O.R.; Foxman, B.M.; Lin, W. Luminescent lanthanide coordination polymers. *Inorg. Chem.* **1999**, *38*, 5837–5840. [CrossRef]

8. Seo, J.S.; Whang, D.; Lee, H.; Jun, S.I.; Oh, J.; Jeon, Y.J.; Kim, K. A homochiral metal-organic porous material for enantioselective separation and catalysis. *Nature* **2000**, *404*, 982–986. [CrossRef]

9. Verlan, V.I.; Iovu, M.S.; Culeac, I.; Nistor, Y.H.; Turta, C.I.; Zubareva, V.E. Photoluminescence properties of PVP/Tb(TTA)$_2$(Ph3PO)$_2$NO$_3$ nanocomposites. *J. Non-Crystalline Solid.* **2013**, *360*, 21–25. [CrossRef]

10. Cui, G.; Chen, S.; Jiang, B.; Zhang, Y.; Qiu, N.; Satoh, T.; Kakuchi, T.; Duan, Q. Synthesis and characterization of novel thermoresponsive fluorescence complexes based on copolymers with rare earth ions. *Optic. Mater.* **2013**, *35*, 2250–2256. [CrossRef]

11. Ye, J.; Li, W.; Xiong, J. Relationship Between Size as Well as Size Distribution and Fluorescence Properties of CMC /Eu Complex Nanoparticles Synthesized at Different pH Values. *J. South China Univ. Technol.* **2014**, *42*, 73–78.

12. Ye, J.; Wang, B.; Xiong, J. Effect of Eu3+ Concentration on the Structure and Fluorescence Quenching of Carboxymethyl Cellulose /Eu(III) Nanoparticles. *Polym. Mater. Sci. Eng.* **2016**, *32*, 32–37.

13. Ye, J.; Li, W.; Xiong, J. Properties of Carboxyl-Containing Cellulose Derivatives/ Rare Earth Ions Nanocomposites. *J. South China Univ. Technol.* **2017**, *45*, 48–56.

14. Ye, J.; Guo, Y.; Xiong, J. Effect of pH on size and dispersion of CMC/Eu nanocomposites with high fluorescence intensity. *Funct. Mater.* **2012**, *18*, 2541–2545.

15. Dai, H.; Wong, E.W.; Lu, Y.Z.; Fan, S.; Lieber, C.M. Synthesis and characterization of carbide nanorods. *Nature* **1995**, *375*, 769–772. [CrossRef]

16. Dong, S.; Roman, M. Fluorescently Labeled Cellulose Nanocrystals for Bioimaging Applications. *J. Am. Chem. Soc.* **2007**, *129*, 13810–13811. [CrossRef] [PubMed]

17. Wu, W.; Huang, F.; Pan, S.; Mu, W.; Meng, X.; Yang, H.; Xu, Z.; Ragauskas, A.J.; Deng, Y. Thermo-responsive and fluorescent cellulose nanocrystals grafted with polymer brushes. *J. Mater. Chem. A* **2015**, *3*, 1995–2005. [CrossRef]

18. Díez, I.; Eronen, P.; Österberg, M.; Linder, M.B.; Ikkala, O.; Ras, R.H.A. Functionalization of Nanofibrillated Cellulose with Silver Nanoclusters: Fluorescence and Antibacterial Activity. *Macromol. Biosci.* **2011**, *11*, 1185–1191. [CrossRef]

19. Niu, Q.; Gao, K.; Wu, W. Cellulose nanofibril based graft conjugated polymer films act as a chemosensor for nitroaromatic. *Carbohydr. Polym.* **2014**, *110*, 47–52. [CrossRef]

20. Xiang, Z.; Jin, X.; Liu, Q.; Chen, Y.; Li, J.; Lu, F. The reinforcement mechanism of bacterial cellulose on paper made from woody and non-woody fiber sources. *Cellulose* **2017**, *24*, 5147–5156. [CrossRef]

21. Xiang, Z.; Liu, Q.; Chen, Y.; Lu, F. Effects of physical and chemical structures of bacterial cellulose on its enhancement to paper physical properties. *Cellulose* **2017**, *24*, 3513–3523. [CrossRef]

22. Zhang, X.; Chen, W.; Lin, Z.; Yao, J.; Tan, S. Preparation and Photocatalysis Properties of Bacterial Cellulose/TiO2Composite Membrane Doped with Rare Earth Elements. *Synth. React. Inorg. Metal-Organic Nano-Metal Chem.* **2011**, *41*, 997–1004. [CrossRef]

23. Yang, Z.; Chen, S.; Hu, W.; Yin, N.; Zhang, W.; Xiang, C.; & Wang, H. Flexible luminescent CdSe/bacterial cellulose nanocomoposite membranes. *Carbohydr. Polym.* **2012**, *88*, 173–178. [CrossRef]

24. Chen, L.F.; Huang, Z.H.; Liang, H.W.; Yao, W.T.; Yu, Z.-Y.; Yu, S.H. Flexible all-solid-state high-power supercapacitor fabricated with nitrogen-doped carbon nanofiber electrode material derived from bacterial cellulose. *Energ. Environ. Sci.* **2013**, *6*, 3331. [CrossRef]

25. Huang, Y.; Wang, T.; Ji, M.; Yang, J.; Zhu, C.; Sun, D. Simple preparation of carbonized bacterial cellulose—Pt composite as a high performance electrocatalyst for direct methanol fuel cells (DMFC). *Mater. Lett.* **2014**, *128*, 93–96. [CrossRef]
26. Kwon, H.; Samain, F.; Kool, E.T. Fluorescent DNAs printed on paper: Sensing food spoilage and ripening in the vapor phase. *Chem. Sci.* **2012**, *3*, 2542. [CrossRef]
27. Ma, Y.; Li, H.; Peng, S.; Wang, L. Highly Selective and Sensitive Fluorescent Paper Sensor for Nitroaromatic Explosive Detection. *Anal. Chem.* **2012**, *84*, 8415–8421. [CrossRef]
28. Su, L.; Yang, L.; Sun, Q.; Zhao, T.; Liu, B.; Jiang, C.; Zhang, Z. A ratiometric fluorescent paper sensor for consecutive color change-based visual determination of blood glucose in serum. *New J. Chem.* **2018**, *42*, 6867–6872. [CrossRef]
29. Zhang, W.; Feng, X.; Zhu, Y.; Wei, X. Study on the Stability and Color Property of Fluorescent Ink-jet Ink. *Lect. Notes Elect. Eng.* **2015**, *369*, 919–925.
30. Hou, X.; Ke, C.; Bruns, C.J.; McGonigal, P.R.; Pettman, R.B.; Stoddart, J.F. Tunable solid-state fluorescent materials for supramolecular encryption. *Nat. Commun.* **2015**, *6*, 6884. [CrossRef]
31. Purington, E.; Bousfield, D.; Gramlich, W.M. Fluorescent dye adsorption in aqueous suspension to produce tagged cellulose nanofibers for visualization on paper. *Cellulose* **2019**, *26*, 5117–5131. [CrossRef] [PubMed]
32. Xiang, Z.; Chen, Y.; Liu, Q.; Lu, F. A highly recyclable dip-catalyst produced from palladium nanoparticle-embedded bacterial cellulose and plant fibers. *Green Chem.* **2018**, *20*, 1085–1094. [CrossRef]
33. Wu, X.; Xiang, Z.; Song, T.; Qi, H. Wet-strength agent improves recyclability of dip-catalyst fabricated from gold nanoparticle-embedded bacterial cellulose and plant fibers. *Cellulose* **2019**, *26*, 3375–3386. [CrossRef]
34. Xu, J.; Huang, X.; Zhou, N.; Zhang, J.; Bao, J.C.; Lu, T.; Li, C. Synthesis, XPS and fluorescence properties of Eu^{3+} complex with polydimethylsiloxane. *Mater. Lett.* **2004**, *58*, 1938–1942. [CrossRef]
35. He, P.; Wang, H.H.; Liu, S.G.; Shi, J.X.; Wang, G.; Gong, M.L. Visible-Light Excitable Europium(III) Complexes with 2,7-Positional Substituted Carbazole Group-Containing Ligands. *Inorg. Chem.* **2009**, *48*, 11382–11387. [CrossRef] [PubMed]

 *nanomaterials*

Article

Arginine/Nanocellulose Membranes for Carbon Capture Applications

Davide Venturi [1], Alexander Chrysanthou [1], Benjamin Dhuiège [2], Karim Missoum [2] and Marco Giacinti Baschetti [1,*]

[1] Department of Civil, Chemical, Environmental and Material Engineering (DICAM), Alma Mater Studiorum, University of Bologna, Via Terracini, 28, 40131 Bologna, Italy; davide.venturi16@unibo.it (D.V.); A.Chrysanthou-15@student.lboro.ac.uk (A.C.)

[2] INOFIB, Rue de la papeterie, 461, 38402 St-Martin-d'Hères, CEDEX, France; benjamin.dhuiege@inofib.com (B.D.); karim.missoum@inofib.com (K.M.)

* Correspondence: marco.giacinti@unibo.it; Tel.: +39-051-209-0408

Received: 14 May 2019; Accepted: 4 June 2019; Published: 10 June 2019

Abstract: The present study investigates the influence of the addition of L-arginine to a matrix of carboxymethylated nanofibrillated cellulose (CMC-NFC), with the aim of fabricating a mobile carrier facilitated transport membrane for the separation of CO_2. Self-standing films were prepared by casting an aqueous suspension containing different amounts of amino acid (15–30–45 wt.%) and CMC-NFC. The permeation properties were assessed in humid conditions (70–98% relative humidity (RH)) at 35 °C for CO_2 and N_2 separately and compared with that of the non-loaded nanocellulose films. Both permeability and ideal selectivity appeared to be improved by the addition of L-arginine, especially when high amino-acid loadings were considered. A seven-fold increment in carbon dioxide permeability was observed between pure CMC-NFC and the 45 wt.% blend (from 29 to 220 Barrer at 94% RH), also paired to a significant increase of ideal selectivity (from 56 to 185). Interestingly, while improving the separation performance, water sorption was not substantially affected by the addition of amino acid, thus confirming that the increased permeability was not related simply to membrane swelling. Overall, the addition of aminated mobile carriers appeared to provide enhanced performances, advancing the state of the art for nanocellulose-based gas separation membranes.

Keywords: CO_2 separation; facilitated transport; nanocellulose; amino acid; gas separation membranes

---

1. Introduction

One of today's major global concerns is represented by the excess of environmental greenhouse gases (GHGs), such as CO_2, in the terrestrial atmosphere [1]. Global warming issues, which are strongly related to these gases, are indeed now recognized by a vast majority of law-makers, and severe actions must be taken in order to minimize emissions and the overall impact of high concentrations of carbon dioxide and other GHGs [2].

Within this frame of thought, technologies such as carbon capture and storage (CCS) appear to play a significant role for short- to medium-term goals in the reduction of anthropogenic CO_2, while a zero-emission path for energy production is built. A CCS approach is based on the removal of CO_2 from flue gases of large-scale energy productions and other industrial activities, and the sequestration of the compressed gas in underground deposits [3,4].

To perform this operation, efficient separation techniques must be adopted in order to guarantee an acceptable degree of purity, while maintaining low costs. Traditionally, carbon dioxide is separated from light gases, such as N_2, CH_4, CO, and H_2, by means of chemical absorption, constantly regenerating the solvent. This methodology is currently quite optimized, but it brings with itself several disadvantages, such as relatively low energy efficiency, difficulty in operability, and the use of harmful organic solvents [5]. Gas separation membranes represent an alternative solution that is gaining an increasing amount of interest in the chemical and process industry, thanks to the absence of additional fluids, reduced energy needs, and absence of moving parts [5–8].

Membranes commonly utilized by the industry rely purely on a solution–diffusion mechanism to act as a molecular sieve substantially based on the kinetic diameter and condensability of the various molecules [9]. This mechanism, however, tends to suffer from an inherent trade-off between permeability and selectivity, which limits the separation performance of the materials [10,11].

A number of strategies were developed over the years to overcome in some way this limitation, using, for example, mixed matrix membranes [12–14] or polymer and block copolymers with high CO_2 affinity [15–18]. Recently, a higher focus was given to facilitated transport membranes (FTM); in this class of materials, the diffusion of a specific molecule (in this case, carbon dioxide) is driven not only by a solution–diffusion mechanism, but also by a reaction of the gas species with functional groups embedded in the matrix, commonly named "carriers" [19–21]. The permeation of a specific gas, or an ensemble of gases with similar chemical characteristics is, therefore, facilitated as they can diffuse both as free molecules or via carrier-mediated transport [22]. This way, since only the desired molecule transport is increased, both overall permeability and selectivity toward other species are enhanced, possibly overcoming the previously described limitations of the solution–diffusion mechanism [23,24].

In the case of carbon dioxide separation, amine moieties represent a fairly reasonable choice which were exploited in several works [25–29]. The mechanism itself, which regulates the interaction between aminated molecules and CO_2, is not completely understood, even though two main reactions are commonly considered to occur within the matrix [29,30]. When unhindered amines are present, carbon dioxide tends to form a carbamate ion through a zwitterion mechanism, a pathway originally described by Caplow [31] and presented in Equations (1) and (2):

$$CO_2 + R - NH_2 \rightleftharpoons R - NH_2^+ - COO^-, \tag{1}$$

$$R - NH_2^+ - COO^- + R - NH_2 \rightleftharpoons R - NH - COO^- + R - NH_3^+. \tag{2}$$

A single molecule of the target gas interacts with two amine groups in order to be facilitated in its diffusion. If a hindered amine is present, a second mechanism is preferred, because of the fact that the carbamate ion has instability issues, due to its large steric hindrance [29]. Hence, the formation of a smaller molecule is favored, such as a hydrogen carbonate ion, as proposed by Kim et al. [28,32] (see Equations (3) and (4)).

$$CO_2 + H_2O \rightleftharpoons H_2CO_3, \tag{3}$$

$$H_2CO_3 + R_1 - NH - R_2 \rightleftharpoons HCO_3^- + R_1 - NH_2^+ - R_2. \tag{4}$$

In this case, the overall stoichiometric ratio between carbon dioxide and a given amine moiety is 1:1, theoretically granting a higher performance with respect to the previously presented mechanism. In both reaction schemes, the mechanisms occurring at the upstream side of the film are outlined, which represent the complexing of CO_2 to the carrier molecule once the gas is dissolved. The opposite reaction will take place at the downstream side, decomplexing the carbon dioxide and allowing the molecule to pass once again to the gaseous phase.

Historically, the earliest examples of FTMs were supported liquid membranes (SLM), consisting of an immobilized liquid phase, where free carriers could freely move and interact with the target species. These would dissolve on the upstream side and form complexes with the mobile carrier; as a unit, this complex would diffuse through the liquid and decomplex on the downstream side [33]. Another approach to this issue is represented by the use of fixed site carrier (FSC) membranes, where the functional groups are covalently bonded to the matrix polymeric backbone or a dispersed secondary phase and, rather than being free to move through the entire volume of it, their position is limited in the vicinity of an equilibrium point [24,30,34]. Both approaches were explored in research and both showed strengths and weaknesses; small mobile carriers tend to leak and evaporate [35], while fixed carriers can struggle to achieve the same diffusion rates due to lack of mobility. For these reasons, this work focused on the usage of mobile carrier molecules with a high vapor tension and low volatility, combined with a strong presence of aminated functional groups. A category of molecules, which checks all the requisites here laid out, is represented by amino acids and amino-acid salts, which are gaining a certain interest in the field of FTMs [36–39].

For this purpose, L-arginine was selected, since it possesses a strong alkaline polar nature, combined with a high solubility in water (182 g/L [40]) and a relatively low price with respect to other amino acids. As a matrix to hold the arginine and allow the uptake of humidity, carboxymethylated nanocellulose (CMC-NFC) was chosen. Cellulose is the most abundant biopolymer on earth. Extracted most often from wood or annual plants, cellulose fibers are composed of several fibrils which were firstly isolated by Turbak et al. [41] in 1983 and are usually present as microfibrillated cellulose (MFC), cellulose nanofibrils (CNF), or nanofibrillated cellulose (NFC). These kinds of nanocelluloses are some of the most studied bio-based materials and are often reported from production until application in several books and reviews [42–45].

Since 2008, NFC development demonstrated exponential interest from researchers, as well as industrial companies, with more than 60 producers worldwide. More recently, in 2016, NFCs were identified as the second bio-economy priority in Europe thanks to their properties. Indeed, NFCs display high mechanical resistance and excellent barrier level, and they are also biodegradable and biocompatible, highlighting this bio-based material as an excellent candidate in several applications such as packaging [44], paper and board [46], composites [47], printed electronics [48], biomedical devices [49], etc. Also, in the field of membranes, nanocellulose is gaining attention; in the last three years, several papers appeared using this material as a base for the production of gas separation membranes [50–53].

In order to isolate NFCs from fibers, an enzymatic or chemical pretreatment is performed in order to weaken cellulosic fibers, as well as to reduce the energy consumption of the production process during the mechanical step [54–57]. One of the chemical pre-treatments applied to cellulosic fibers is carboxymethylation [58,59]. This modification was already reported in numerous papers with specific properties given to carboxymethylated NFC, i.e., highly charged surface (which can lead to ionic interactions with other molecules), and a lot of acid groups at the surface (for post-modification purpose) in comparison to neat NFC [60,61]. Due to its highly polar surface, CMC-NFC fibers were deemed to have a higher chance than plain nanocellulose to positively interact with the amino acid arginine, which is rich in alkaline polar groups. Overall, this combination represents an interesting approach to gas separation membranes and the use of sustainable materials.

Following this line of thought, in the present work, different CMC-NFC/L-arginine composite films were prepared and tested to understand their potential as CO_2 separation membranes. Permeability tests were performed for CO_2 and N_2 at 35 °C and different relative humidity; water vapor sorption experiments were also conducted at the same temperature to relate the permeability to actual water content in the membrane and to better understand its role in the enhancement of membrane performance.

As stated, carboxymethylated NFC can easily react with different molecules due to its amine functional group. In the line of sustainability and green technology development, it appears essential to prepare new membranes for carbon capture using natural molecules, i.e., arginine and bio-based membranes such as CMC-NFC. To the best of our knowledge, there are very few papers and amine-based functional groups in carbon capture membranes, mainly dealing with NFC - aminated polymers blends [50,51] or with neat NFC grafted with aminosilanes [62]. In the present work, CMC-NFC was used as a membrane with arginine as a mobile carrier for CO_2.

2. Materials and Methods

2.1. Carboxymethylated Nanocellulose Synthesis

Carboxymethylated nanocellulose (CMC-NFC) was kindly provided by INOFIB (Saint-Martin-d'Hères Cedex, France) as a water suspension with a solid content of 1.7 wt.% and a surface charge of 2600 µequiv/mol.

CMC is usually synthetized by the alkali-catalyzed reaction of cellulose with chloroacetic acid. The protocol used for the modification was based on the one developed by Wågberg et al. [58]

A total of 180 g of eucalyptus fiber was pretreated. The eucalyptus fibers were first dispersed in water and then solvent-changed to 4 L of ethanol. The fibers were then impregnated with a solution of 175.1 g of monochloroacetic acid in 820 mL of isopropanol corresponding to 10 M OH equivalent with respect to cellulose. A solution of 26.5 g of NaOH in 3.3 L of isopropanol was then added to fibers that were heated to just below boiling temperature in a 10-L reactor fitted with a condenser. This carboxymethylation reaction was allowed to continue for 6 h. Following this carboxymethylation step, the fibers were filtered and washed: first with 20 L of deionized water, then with 3 L of acetic acid (0.1 M), and finally with 15 L of deionized water. The fibers were then impregnated with a 3-L $NaHCO_3$ solution (4 wt.% solution) for 60 min in order to convert the carboxyl groups to their sodium form. Finally, the fibers were washed with 15 L of deionized water and drained on a Buchner funnel. In order to obtain CMF-NFC, carboxymethylated fibers were mechanically treated using a Masuko Grinder® device with a speed of 1500 rpm and a gap of −10 µm between the two grinding stones. A gel at 1.7 wt.%. CMC-NFC was obtained after 10 passes.

Morphological studies on CMC-NFC were performed using the FEI-Quanta 200 Scanning Electron Microscope (SEM). The accelerating voltage (Extra High Tension or EHT) was 10 kV for a working distance of 9.7 mm. A drop of diluted CMC-CNF suspension was deposited onto a substrate covered with carbon tape and dried using a vacuum pump and then coated with a layer of Au/Pd (gold/palladium) and an Everhart–Thornley Detector (EDT) was used.

The CMC-NFC gel was characterized by conductimetric titration according to the ISO 638:2008 procedure. The CMC-NFC was acidified to a pH around 2.8 using hydrochloric acid (0.1 M) to convert the carboxyl groups into their acid form (–COOH). The titration was performed with a sodium hydroxide solution (C = 0.05 M). The conductivity of the suspension was measured for each addition of NaOH solution, until pH 10 was reached. Samples were triplicated and the average of the equilibrium concentration was used for calculation of charge.

Infrared spectra were recorded on films on CMC-NFC, using a Perkin-Elmer SP100 spectrometer. For each sample, the diamond crystal of an attenuate total reflectance (ATR) apparatus was used. The torque applied was kept constant to ensure the same pressure on each sample. Triplicates were performed for each sample and the best representative spectra were kept for consideration. All spectra were recorded between 4000 and 600 cm^{-1}, with a resolution of 1 cm^{-1} and eight scans.

2.2. Membrane Fabrication

A solvent casting protocol, similar to previous works [50,51], was adopted, in order to obtain homogeneous films. L-Arginine (purchased from Sigma-Aldrich with a reported purity >99.5%) was added directly to the CMC-NFC water suspension in powder form and dissolved at room temperature via magnetic stirring for 1 h at 1000 rpm, until no solids could be discerned in the solution. Three different loadings of the amino acid were investigated (15, 30, and 45 wt.%, calculated with respect to the solid content of the suspension) plus a blank sample of pure CMC-NFC.

Due to its high viscosity, during the stirring phase, the suspension tended to incorporate a large amount of air in the form of bubbles. These would eventually result in imperfections in the final film. For this reason, it was centrifuged in mild conditions (10 min, 4500 rpm), in order to separate the gas phase from the liquid one. Subsequently, the homogenized blend was poured into glass petri dishes (11 cm in diameter) and dried in a ventilated oven at 35 °C on a leveled plane for two days. Once dried, the films were peeled off and their thickness was evaluated via a disc micrometer (Mitutoyo, Series 227-221). The thickness was between 35 and 45 μm for the arginine-loaded films and between 15 and 20 μm for the unloaded membrane. An average of three samples were fabricated for each composition.

2.3. Water Sorption

Water uptake was tested for the different samples in film form via a quartz spring microbalance [63]. The basic functioning of this system relies on a quartz spring of known elastic constant, from the bottom of which the sample is hung. Both spring and specimen are enclosed in a glass, thermostatic column, where the pressure of water vapor can be controlled (Figure 1).

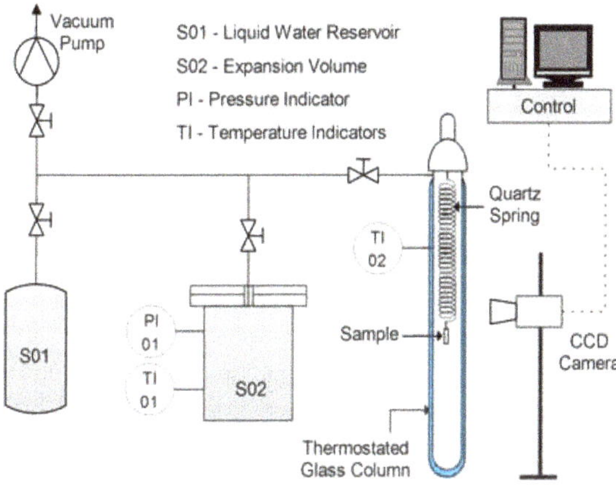

Figure 1. Layout of the quartz spring microbalance used to measure water vapor uptake.

The variation in weight of the sample is recorded by measuring its displacement via a digital charge-coupled device (CCD) camera. The amount of water gained by the sample at each step can be obtained by applying Equation (5).

$$[m_{water}]_i = (h_{i,t\to\infty} - h_0)\cdot\frac{k}{g}, \tag{5}$$

where h represents the vertical coordinate of the sample, k is the elastic constant of the spring, and g is the gravitational acceleration. Moreover, through the analysis of the transitory phase of a sorption step, the diffusion coefficient of water in the film can also be estimated at a given concentration via Equation (6) [64].

$$\frac{m_{sample}}{m_{sample,\ t\to\infty}} = 1 - \sum_n \frac{8}{(2n+1)^2\pi^2}exp\left[\frac{-\mathcal{D}(2n+1)^2\pi^2 t}{L^2}\right]. \tag{6}$$

The diffusion coefficient D is fitted via a graphical interpolation of the experimental data as a function of time t, knowing the half-thickness of the sample (L).

All tests were carried out at a temperature of 35 °C and with a water activity ranging from 0.25 to 0.80.

2.4. Permeability

Single-gas permeability in humid conditions of the fabricated films was evaluated for two gases, CO_2 and N_2. Before testing, the samples were prepared by cutting them in a circular shape and masking the outer part with aluminum tape sealed with epoxy resin. This was done to avoid direct compression of the film by the rubber O-ring utilized to ensure a good seal in the sample holder. The apparatus used for this purpose, the layout of which is presented in Figure 2, had a set-up incorporating a fixed volume, variable pressure, and a humid permeometer.

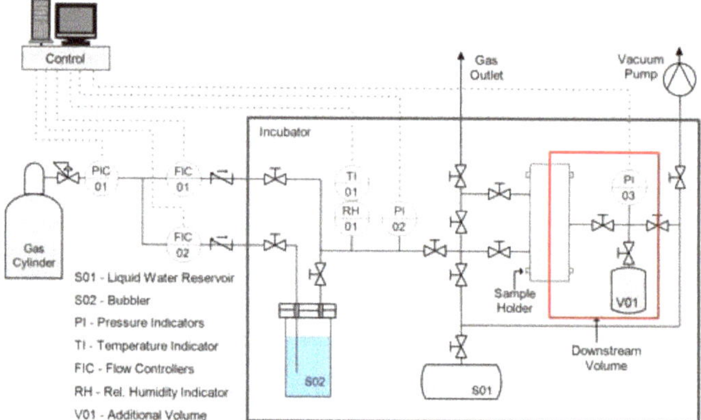

Figure 2. Diagram of the humid single-gas permeation apparatus.

A thorough description of the system and its corresponding protocol can be found in a previous work [65]. Pressure (p_1) is constantly measured and recorded in the downstream volume (V). On the upstream side, pressure p_2 is maintained constant, while a flow of humidified gas is fed to the sample. Given these two values and the variation of the downstream pressure over time, Equation (7) can be applied to calculate permeability, assuming an ideal gas state.

$$P = -\left(\frac{dp_1}{dt}\right)_{t\to\infty} \frac{V}{RT}\frac{L}{A}\frac{1}{(p_1 - p_2)}, \tag{7}$$

where A represents the effective area of the membrane, L is its thickness, T is the absolute temperature, and R is the ideal gas constant.

Prior to experiments, the sample and the system were evacuated overnight via a vacuum pump, in order to remove all volatile compounds such as gases absorbed by the film or compounds from the adhesives used in the sample preparation. The following step consisted of introducing pure water vapor to condition the film prior the permeation; pressure was controlled until it reached a stationary value, corresponding to the relative humidity, at which the test was conducted. The gas upstream was then humidified to the same value of the conditioning step, and the test was started by exposing the membrane to the gas. With water in a condition of thermodynamic equilibrium between the two sides of the film, the only contribution to the pressure increment in the downstream volume was eventually represented by the permeation of the incondensable gas (either carbon dioxide or nitrogen, depending on the test). Throughout this work, permeability is expressed in Barrer, and its definition and conversion to International System of Units (SI) units are outlined in Equation (8).

$$1\ \text{Barrer} = 10^{-10}\frac{\text{cm}^3(\text{STP})\cdot\text{cm}}{\text{cm}^2\cdot\text{cmHg}\cdot\text{s}} = 3.364\cdot 10^{-16}\frac{\text{mol}}{\text{m}\cdot\text{Pa}\cdot\text{s}}. \tag{8}$$

Gas selectivity was evaluated as the ideal ratio between pure gas permeability, an approximation justified by the low downstream pressure. Tests were performed at 35 °C, with a humidity ranging between 70% and 95%, as lower values were generally characterized by very low permeability.

3. Results and Discussion

3.1. CMC-NFC Characterization

Firstly, a visual observation after the grinding process is shown in Figure 3. This picture clearly shows the gel-like structure and the transparency of the films obtained with the pretreatment done on the fibers.

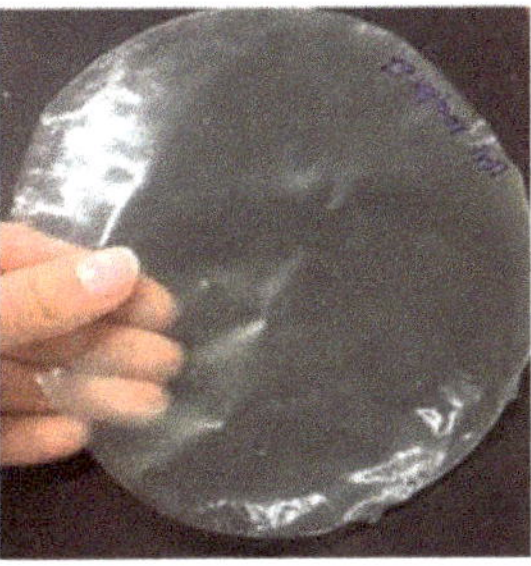

Figure 3. Images of the carboxymethylated nanofibrillated cellulose (CMC-NFC) suspension (**left**) and a sample of the casted film (**right**).

Moreover, SEM analyses were performed to point out the nanostructure of CMC-NFC, as represented in Figure 4.

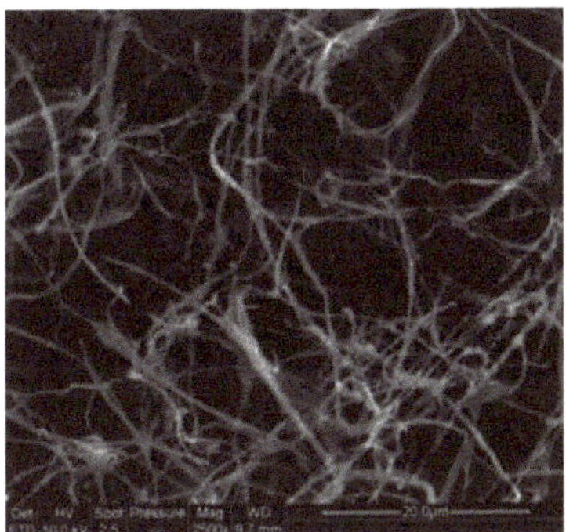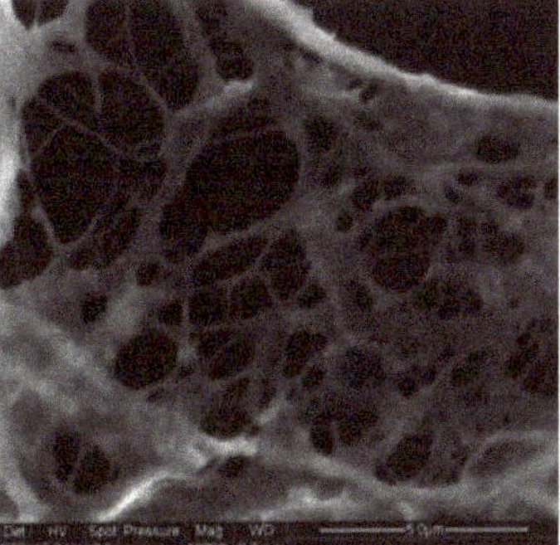

Figure 4. SEM images of a film of carboxymethylated nanocellulose, highlighting the fibrous three-dimensional structure of the material.

The diameter of CMC-NFC was between 80 nm and 150 nm, determined by digital image analysis (ImageJ software) of SEM pictures (a minimum of 50 measurements was performed). Figure 5 presents the quantification of carboxylic content following the titration method from ISO 638:2008.

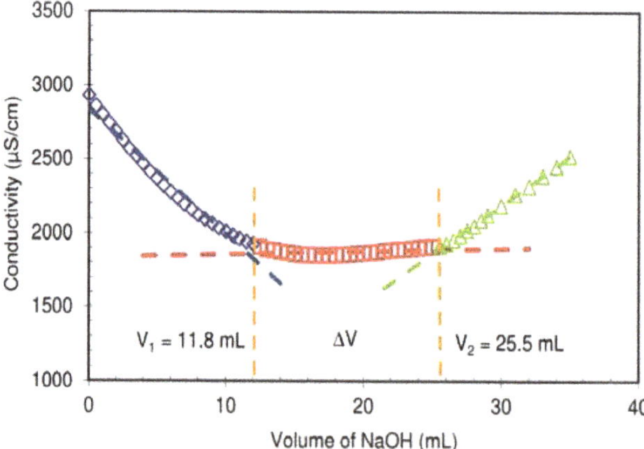

Figure 5. Outcome of conductimetric titration on a CMC-NFC suspension for determining the surface charge of the nanofibers.

In comparison to the cellulosic fibers used as raw materials for the preparation of these NFCs, displaying a total charge around 22 $\mu eq \cdot g^{-1}$ (data not shown) mainly originating from hemicelluloses, Figure 5 clearly shows that CMC-NFC was charged thanks to carboxymethylation pre-treatment yielding the formation of –COOH groups which cause the wide plateau visible in the titration chart. The total charge of CMC-NFC was around 2600 $\mu eq \cdot g^{-1}$ of carboxylic groups and confirms the occurrence of the modification.

To ensure this first conclusion, Fourier-transform infrared (FTIR) spectroscopy was also performed as shown in Figure 6, where the FTIR spectra of both the cellulose fiber sample (dotted line) and the CMC-NFC (full line) are presented. They display similar characteristic bands attributed to cellulose substrates. Indeed, the bands around 3496 cm^{-1} (O–H), 1110 cm^{-1} (C–O of secondary alcohol) (used for the normalization of spectra), and 2868 and 2970 cm^{-1} (C–H from –CH$_2$–) are present in both samples.

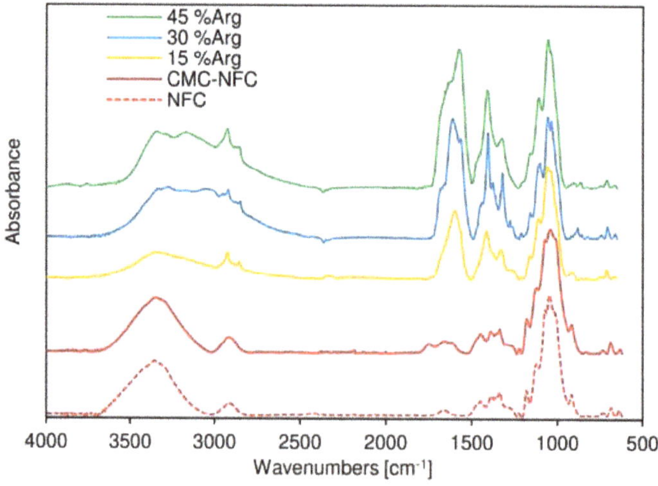

Figure 6. Fourier-transform infrared (FTIR) spectra of plain cellulose, carboxymethylated nanocellulose, and the CMC-NFC/arginine blends.

The main difference between the two spectra was the peak attributed to carboxylic acid vibration around 1745 cm^{-1} due to the larger number of carboxylic groups provided by the carboxymethylation process. A second peak can be also attributed to the ionized form of the carboxylic acid group around 1614 cm^{-1}.

FTIR spectra of the CMC-NFC/arginine blends were also acquired in order to assess the presence of the amino acid within the polymeric matrix. A new peak was first encountered at 3151 cm^{-1}, related to the stretching of the N–H bond which extended and partially overlapped with the O–H stretching band. The resulting broad peak suggested the presence of an extensive hydrogen bonding network in the blend. Another significant indication of the growing presence of L-arginine was determined by the sharp peak at 1590 cm^{-1}, which was due to the out-of-plane bending of the N–H group, and the series of peaks between 1420 and 1320 cm^{-1} could be related to the symmetrical bending of the CH$_3$ group, as suggested by Kumar et al. [66].

In conclusion, FTIR analysis confirmed that plain cellulose was successfully modified into carboxymethylated nanocellulose and efficiently blended with different amounts of arginine.

3.2. Water Sorption

All the isothermal sorption curves are collected in Figure 7, which presents the water uptake expressed with respect to the dry weight of the film as a function of water activity. From a first view of the data, no significant difference can be outlined when the content of amino acid was incremented in the film. This shows that the water uptake of the nanocellulose/arginine blend does not seem to depend on the ratio between the two components, even if some difference can be observed in the general behavior. In particular, up to an activity of 0.5 and an uptake of about 0.1 g/g$_{pol}$, the water concentration increased with an approximately linear slope with pure NFC films, showing higher water uptake and a positive intercept of the linear trend with respect to the loaded samples. From this point onward, the slope of the isothermal curve changed, acquiring a more exponential trend. Values here ranged between 0.15 g/g$_{pol}$ at an activity of 0.60, up to 0.33 g/g$_{pol}$ at 0.79, and all curves were superimposed with no differences visible among different materials. This dual trend was already observed in previous works on vapor sorption in NFC [50,51], and it was explained by a first phase controlled by the adsorption of water onto the polar moieties of the fibers, followed by a second phase commonly linked to water clustering [67,68] and film swelling, which is also related to the adsorption of a subsequent layer of water molecules onto the first one, exponentially increasing the uptake of water.

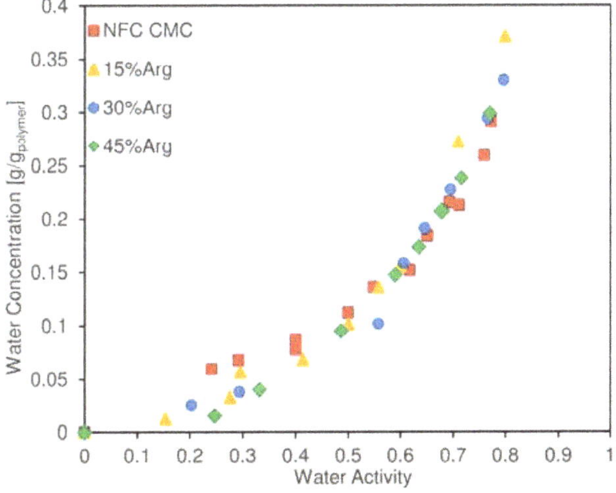

Figure 7. Water uptake of CMC-NFC films with different loadings of arginine at 35 °C.

Considering the differences between pure and arginine-loaded NFC, it seems that, at low RH, the pure cellulose nanofibrils adsorb water more easily than composite films, suggesting that, in the latters, some of the acid groups on the surface are less available, due very likely to the interaction with the alkaline part of the amino acid. Interestingly, the difference indeed disappears at higher humidity when the adsorption of subsequent water layers or absorption in the interfibrillar space is considered.

In general, the absence of any difference among different films with very different loadings of arginine suggests a similar overall water uptake of both NFC fibers and the amino acid, thereby not varying the total equilibrium water concentration while the relative amount of the two changed.

Another piece of information acquired from quartz spring balance tests involved an estimation of the diffusion coefficient of water in the hydrated matrix, which is presented in Figure 8 as obtained from the use of Equation (6) on the experimental data of weight change over time.

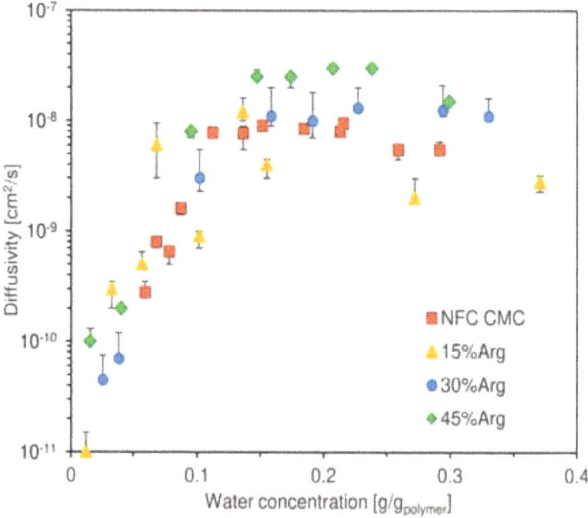

Figure 8. Fickian diffusion coefficient of CMC-NFC films with different loadings of arginine at 35 °C.

The reported values represent, therefore, the result of a graphical interpolation, assuming a purely Fickian behavior of the diffusion mechanic. As per the water uptake, it is difficult to discern a significant difference between the curves corresponding to the different loadings of arginine. Again, similarly to that observed in other works regarding nanocellulosic membranes [51,52], diffusivity increased exponentially at low water uptake, specifically between a concentration of 0 and 0.1 g/g_{pol}. From the lowest value measured (1.5×10^{-11} cm^2/s at 0.01 g/g_{pol}), diffusivity increased by three orders of magnitude reaching 1.2×10^{-8} cm^2/s at 0.13 g/g_{pol}. From a concentration of 0.1 g/g_{pol}, the exponential trend stopped, and a quite stable plateau was reached for all, with values ranging from 3×10^{-9} to 3×10^{-8} cm^2/s (as a reference, the self-diffusion coefficient of water in a liquid state at 35 °C is 2.9×10^{-5} cm^2/s [69]).

In this part of the chart, it could be argued that the different samples tended to reach a plateau at different asymptotic values, for the film with 45 wt.% Arginine occupied the top place at 3×10^{-8} cm^2/s; however, due to the data scattering, high experimental uncertainty was associated with different data. Thus, a final conclusion on the presence of a trend in the asymptotic diffusivity cannot be given without a deeper analysis, which is outside the scope of the present work.

In general, the qualitative trend of diffusivity was justified with the two phases of water sorption, outlined earlier in this section. At low water concentration, the molecules had to diffuse through a tightly packed matrix, determining very low diffusivity values. This process continued until a hydration of about 0.1 g/g_{pol}; from this point onward, the nanocellulosic matrix appeared to absorb enough humidity to cause a generalized swelling, which allowed water to diffuse with minimal interaction with the nanocellulosic fibers, thus presenting improved mobility. At the highest water uptakes investigated, diffusivity appeared to show the onset of a downward trend; this would be in accordance with that previously observed for cellulosic fibers by Gouanvé et al. [70].

3.3. Permeability

In Figure 9, the permeation data for CO_2 and N_2 are presented, collected for each loading of arginine investigated and expressed as a function of relative humidity. As outlined in previous works [12,50,52,71], the trend of gas permeability with humidity in such hydrophilic materials is an exponential one, with variations of up to 2–3 orders of magnitude going from dry to fully humidified samples. For this reason, the investigation involved a water activity not lower than 0.6, since, below this value, permeability of CO_2 reached too low fluxes to be interesting for gas separation applications. In the case of the sample with 45% arginine, the investigated RH range was reduced further due to the difficulties working with this material. The very high amount of amino acid indeed caused a loss of mechanical stability in the film, which made it very difficult to obtain reliable data at RH lower than 80%.

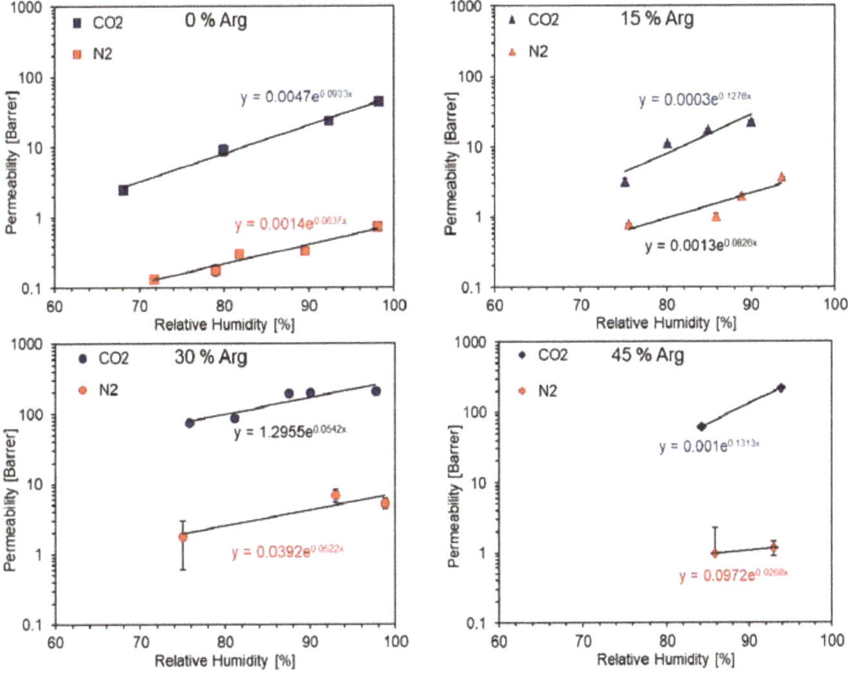

Figure 9. Single-gas permeation results of CMC-NFC with different loadings of arginine with respect to relative humidity.

As a reference, pure CMC-NCF showed a CO_2 permeability ranging from 2.6 Barrer at 68% relative humidity to 45 Barrer at 98.2% RH. These values are in line with that previously observed in similar conditions for unmodified NFC [51].

As expected, nitrogen permeability resulted constantly significantly lower than that of carbon dioxide, passing from 0.13 to 0.75 Barrer when RH was increased from 72% to 98%. For the membrane with a content of arginine of 15 wt.%, CO_2 permeability did not appear to be immediately affected. Indeed, for this membrane, the measured permeability was between 3.2 and 23 Barrer (for a relative humidity range from 75% to 90%). As a consequence, CO_2 permeation did not display a significant variation regardless of the RH in comparison to the pure NFC reference. N_2 permeation, instead, resulted higher than pure CMC-NFC, with values from 0.8 to 3.6 Barrer, thus causing a general decrease in the separation performance of the material.

Interestingly, membranes containing a higher amount of arginine showed a different behavior with respect to previous ones. CO_2 permeability indeed presented a significant increment, with respect to that measured for the non-loaded film with a lower influence of RH. Indeed, for 30 wt.% Arginine loading, values ranged from 75 Barrer at 76% RH to 208 Barrer at 98% RH, while, at 45 wt.%, a CO_2 permeability of 62 Barrer was measured at 84% RH, reaching a value of 220 Barrer when RH was increased to 94%.

In terms of N_2 permeation, values obtained at high arginine loading were also higher than those observed for pure CMC-NFC. The increase was, however, lower than in the previous case, thus leading to an overall increase in the membrane separation performance, as better explained in the following section.

Considering that CO_2 permeability increased at high arginine loading, it is interesting to note that it happened without changing the overall water content in the matrix. This is very different to what was, for example, noted in a previous work [50], where the increment observed when a hydrophilic, aminated polymer was added to NFC resulted indeed very closely related to the increased water uptake of the material. In that case, therefore, the higher swelling induced by water seemed to be the main reason for the improved separation performance, leading to the secondary effect of facilitated transport. Following this thought, the present outcome could be seen as a significant indication (but not definitive proof) of the presence of facilitated transport within the membranes tested.

3.4. Selectivity

Ideal selectivity was calculated as the ratio between permeability of CO_2 and N_2 measured at the same relative humidity. Since RH values differed slightly between points taken for the two gases, an interpolation was adopted. Standard humidity levels were fixed, and the respective permeability values were calculated assuming an exponential trend ($P = Aexp(B\,RH)$), which for sake of clarity is reported in Figure 9 together with the equation used for calculations.

In Figure 10, results of the described procedure are reported in a Robeson plot, where experimental data are presented together with the indication of the relative humidity considered for calculation. In the chart, CMC-NFC clearly showed an incremental trend of ideal selectivity, along with permeability, which was also observed in a previous work [52] for a similar material. In particular, selectivity values ranged from 25 to 56, which represented good separation factors, but still coupled with relatively low permeabilities. Once a small amount of amino acid was introduced (15 wt.%), selectivity values appeared to decrease quite significantly, varying from 4–10 as already expected from the analysis of CO_2 and N_2 selectivity variation.

Once the concentration of L-arginine reached 30 wt.%, selectivity was calculated to range between 30 and 40, without a notable trend with either CO_2 permeability or relative humidity. These values were quite close to pure NFC but coupled with a significantly higher carbon dioxide permeability (maxed at over 200 Barrer), determining an overall improvement with respect to the base material. Finally, the highest content of amino acid (45 wt.%) determined a quite significant increment in selectivity of CO_2 with respect to N_2, showing values from 73 to over 180. In particular, this last point, achieved at very high relative humidity (94 RH%), showed both high permeability and selectivity and placed itself well above Robeson's upper bound [10], confirming the material's potential as a membrane for CO_2 purification.

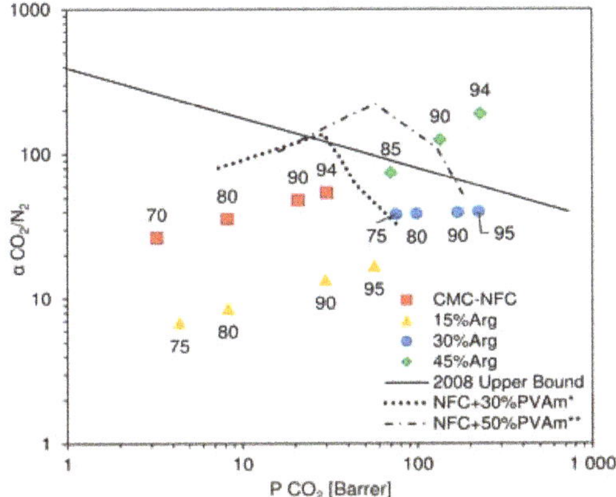

Figure 10. Ideal selectivity of NFC with various contents of L-arginine compared to Robeson's upper bound and NFC/Polyvinylamine blends from previous works [50,51].

As it can also be inferred from the permeation data, the increment in selectivity measured for high loadings of arginine appeared to be mostly related to an improved CO_2 permeability. The fluxes of this gas indeed increased more than the ones of N_2 at all degrees of humidity investigated, leading to a separation factor continuously improving with water content in the membrane. Interestingly, the present behavior is once again in contrast to that observed when coupling nanocellulose with polyvinyl amine. In that case, the data, also reported in Figure 10 for the sake of completeness, showed a well-defined maximum in selectivity at intermediate RH, before decreasing to values rather close to those which could be calculated for a theoretical water membrane. Also, in this case, the different behavior strongly suggests the existence of facilitated CO_2 transport across the membrane, which dominates permeability changes in the whole RH range inspected.

4. Conclusions

Carboxymethylated nanocellulose was synthesized starting from natural fibers and was successfully blended with different quantities of the amino acid L-arginine, acting as a mobile carrier for the transport facilitation of carbon dioxide.

Several self-standing films were casted from the blended suspensions and tested for water sorption and gas permeability in humid conditions.

The addition of L-arginine greatly improved both the permeability and ideal selectivity with respect to a pure carboxymethyl nanocellulose matrix. CO_2 permeability increased seven-fold compared to the base material, from 29 to 225 Barrer, and selectivity with respect to N_2 grew from 55 to 187 when the matrix was loaded with 45 wt.% amino acid.

Moreover, water uptake experiments allowed indicating that the increment in transport properties was not related to the higher concentration of water in the swollen film, as water solubility was not substantially affected by arginine loading. This is a strong indication, but not definitive proof, of the presence of a facilitated transport mechanism, catalyzed by the presence of the mobile carrier.

While further work is needed to study the stability and durability of the materials, the present results are surely of interest and confirm the high potential of nanocellulose-based facilitated transport membranes for carbon capture application.

Author Contributions: Data curation, D.V.; investigation, D.V., A.C., and K.M.; supervision, M.G.B. writing—original draft, D.V.; writing—review and editing, K.M, M.G.B., B.D.

Funding: This research was funded by the European Commission within the NanoMEMC[2] project. This project received funding from the European Union's Horizon 2020 Research and Innovation program under Grant Agreement n° 727734.

Acknowledgments: The authors would like to acknowledge the help of Lorenzo Lucchini in obtaining the water sorption and diffusivity data.

Conflicts of Interest: The authors declare no conflicts of interest.

References

1. Earth System Research Laboratory Global Monitoring Division. Available online: https://www.esrl.noaa.gov/gmd/ccgg/trends/full.html (accessed on 8 May 2019).
2. Masson-Delmotte, V.; Zhai, P.; Pörtner, H.O.; Roberts, D.; Skea, J.; Shukla, P.R.; Pirani, A.; Moufouma-Okia, W.; Péan, C.; Pidcock, R.; et al. *Global Warming of 1.5 °C an IPCC Special Report*; IPCC: Geneva, Switzerland, 2013. [CrossRef]
3. Aaron, D.; Tsouris, C. Separation of CO_2 from Flue Gas: A Review. *Sep. Sci. Technol.* **2005**, *40*, 321–348. [CrossRef]
4. D'Alessandro, D.M.; Smit, B. Carbon Dioxide Capture. *Angew. Chem. Int. Ed.* **2010**, *49*, 6058–6082. [CrossRef]
5. Baker, R.W. Future Directions of Membrane Gas Separation Technology. *Ind. Eng. Chem. Res.* **2002**, *41*, 1393–1411. [CrossRef]
6. Favre, E. Carbon dioxide recovery from post-combustion processes: Can gas permeation membranes compete with absorption? *J. Membr. Sci.* **2007**, *294*, 50–59. [CrossRef]
7. Bernardo, P.; Drioli, E.; Golemme, G. Membrane gas separation: 1 review of state of the art. *Ind. Chem. Eng.* **2009**, *48*, 4638–4663. [CrossRef]
8. Khalilpour, R.; Mumford, K.; Zhai, H.; Abbas, A.; Stevens, G.; Rubin, E.S. Membrane-based carbon capture from flue gas: A review. *J. Clean. Prod.* **2015**, *103*, 286–300. [CrossRef]
9. Baker, R.W.; Lokhandwala, K. Natural Gas Processing with Membranes: An Overview. *Ind. Eng. Chem. Res.* **2008**, *47*, 2109–2121. [CrossRef]
10. Robeson, L.M. The upper bound revisited. *J. Membr. Sci.* **2008**, *320*, 390–400. [CrossRef]
11. Wijmans, J.G.; Baker, R.W. The Solution-Diffusion Model—A Review. *J. Membr. Sci.* **1995**, *107*, 1–21. [CrossRef]
12. Fernández-Barquín, A.; Rea, R.; Venturi, D.; Giacinti-Baschetti, M.; De Angelis, M.G.; Casado-Coterillo, C.; Irabien, Á. Effect of relative humidity on the gas transport properties of zeolite A/PTMSP mixed matrix membranes. *RSC Adv.* **2018**, *8*, 3536–3546. [CrossRef]
13. Rezakazemi, M.; Ebadi Amooghin, A.; Montazer-Rahmati, M.M.; Ismail, A.F.; Matsuura, T. State-of-the-art membrane based CO_2 separation using mixed matrix membranes (MMMs): An overview on current status and future directions. *Prog. Polym. Sci.* **2014**, *39*, 817–861. [CrossRef]
14. Mahmoudi, A.; Asghari, M.; Zargar, V. CO_2/CH_4 separation through a novel commercializable three-phase PEBA/PEG/NaX nanocomposite membrane. *J. Ind. Eng. Chem.* **2015**, *23*, 238–242. [CrossRef]
15. Chen, Y.; Wang, B.; Zhao, L.; Dutta, P.; Winston Ho, W.S. New Pebax®/zeolite Y composite membranes for CO_2 capture from flue gas. *J. Membr. Sci.* **2015**, *495*, 415–423. [CrossRef]
16. Nafisi, V.; Hägg, M.B. Development of dual layer of ZIF-8/PEBAX-2533 mixed matrix membrane for CO_2 capture. *J. Membr. Sci.* **2014**, *459*, 244–255. [CrossRef]
17. Scholes, C.A.; Bacus, J.; Chen, G.Q.; Tao, W.X.; Li, G.; Qader, A.; Stevens, G.W.; Kentish, S.E. Pilot plant performance of rubbery polymeric membranes for carbon dioxide separation from syngas. *J. Membr. Sci.* **2012**, *389*, 470–477. [CrossRef]
18. Tena, A.; Shishatskiy, S.; Filiz, V. Poly(ether-amide) vs. poly(ether-imide) copolymers for post-combustion membrane separation processes. *RSC Adv.* **2015**, *5*, 22310–22318. [CrossRef]
19. Noble, R.D. Generalized microscopic mechanism of facilitated transport in fixed site carrier membranes. *J. Membr. Sci.* **1992**, *75*, 121–129. [CrossRef]
20. Ward, W.J.; Robb, W.L. Carbon dioxide-oxygen separation: Facilitated transport of carbon dioxide across a liquid film. *Science* **1967**, *156*, 1481–1484. Available online: http://www.scopus.com/inward/record.url?eid=2-s2.0-0014208271&partnerID=tZOtx3y1 (accessed on 24 April 2019). [CrossRef]
21. Huang, J.; Zou, J.; Ho, W.S.W. Carbon Dioxide Capture Using a CO_2—Selective Facilitated Transport Membrane. *Ind. Eng. Chem. Res.* **2008**, *47*, 1261–1267. [CrossRef]
22. Rea, R.; De Angelis, M.G.; Baschetti, M.G. Models for facilitated transport membranes: A review. *Membranes* **2019**, *9*, 26. [CrossRef]

23. Matsuyama, H.; Teramoto, M.; Sakakura, H.; Iwai, K. Facilitated transport of CO_2 through various ion exchange membranes prepared by plasma graft polymerization. *J. Membr. Sci.* **1996**, *117*, 251–260. [CrossRef]

24. Matsuyama, H.; Terada, A.; Nakagawara, T.; Kitamura, Y.; Teramoto, M. Facilitated transport of CO_2 through polyethylenimine/poly(vinyl alcohol) blend membrane. *J. Membr. Sci.* **1999**, *163*, 221–227. [CrossRef]

25. Teramoto, M.; Huang, Q.; Maki, T.; Matsuyama, H. Facilitated transport of SO_2 through supported liquid membrane using water as a carrier. *Sep. Purif. Technol.* **1999**, *16*, 109–118. [CrossRef]

26. Matsuyama, H.; Teramoto, M.; Sakakura, H. Selective permeation of CO_2 through poly 2-(*N,N*-dimethyl)aminoethyl methacrylate membrane prepared by plasma-graft polymerization technique. *J. Membr. Sci.* **1996**, *114*, 193–200. [CrossRef]

27. Deng, L.; Kim, T.J.; Hägg, M.B. Facilitated transport of CO_2 in novel PVAm/PVA blend membrane. *J. Membr. Sci.* **2009**, *340*, 154–163. [CrossRef]

28. Kim, T.J.; Baoan, L.I.; Hägg, M.B. Novel fixed-site-carrier polyvinylamine membrane for carbon dioxide capture. *J. Polym. Sci. Part B Polym. Phys.* **2004**, *42*, 4326–4336. [CrossRef]

29. Zhao, Y.; Winston Ho, W.S. Steric hindrance effect on amine demonstrated in solid polymer membranes for CO_2 transport. *J. Membr. Sci.* **2012**, *415*, 132–138. [CrossRef]

30. Tong, Z.; Vakharia, V.K.; Gasda, M.; Ho, W.S.W. Water vapor and CO_2 transport through amine-containing facilitated transport membranes. *React. Funct. Polym.* **2015**, *86*, 111–116. [CrossRef]

31. Caplow, M. Kinetics of carbamate formation and breakdown. *J. Am. Chem. Soc.* **1968**, *90*, 6795–6803. [CrossRef]

32. Sandru, M.; Kim, T.J.; Hägg, M.B. High molecular fixed-site-carrier PVAm membrane for CO_2 capture. *Desalination.* **2009**, *240*, 298–300. [CrossRef]

33. LeBlanc, O.H., Jr.; Ward, W.J.; Matson, S.L.; Kimura, S.G. Facilitated transport in ion-exchange membranes. *J. Memb. Sci.* **1980**, *6*, 339–343. [CrossRef]

34. Li, F.; Li, Y.; Chung, T.-S.; Kawi, S. Facilitated transport by hybrid POSS®–Matrimid®–Zn^{2+} nanocomposite membranes for the separation of natural gas. *J. Membr. Sci.* **2010**, *356*, 14–21. [CrossRef]

35. Kovvali, A.S.; Sirkar, K.K. Dendrimer Liquid Membranes: CO_2 Separation from Gas Mixtures. *Ind. Eng. Chem. Res.* **2001**, *40*, 2502–2511. [CrossRef]

36. Zhang, H.; Tian, H.; Zhang, J.; Guo, R.; Li, X. Facilitated transport membranes with an amino acid salt for highly efficient CO_2 separation. *Int. J. Greenh. Gas Control.* **2018**, *78*, 85–93. [CrossRef]

37. El-Azzami, L.A.; Grulke, E.A. Carbon dioxide separation from hydrogen and nitrogen facilitated transport in arginine salt-chitosan membranes. *J. Membr. Sci.* **2009**, *328*, 15–22. [CrossRef]

38. Moghadam, F.; Kamio, E.; Yoshizumi, A.; Matsuyama, H. An amino acid ionic liquid-based tough ion gel membrane for CO_2 capture. *Chem. Commun.* **2015**, *51*, 13658–13661. [CrossRef]

39. Kasahara, S.; Kamio, E.; Yoshizumi, A.; Matsuyama, H. Polymeric ion-gels containing an amino acid ionic liquid for facilitated CO_2 transport media. *Chem. Commun.* **2014**, *50*, 2996–2999. [CrossRef]

40. National Center for Biotechnology Information. PubChem Database. Arginine, CID=6322. Available online: https://pubchem.ncbi.nlm.nih.gov/compound/6322 (accessed on 30 April 2019).

41. Turbak, A.F.; Snyder, F.W.; Sandberg, K.R. Microfibrillated cellulose, a new cellulose product: Properties, uses, and commercial potential. In *J. Appl. Polym. Sci. Appl. Polym. Symp.*; ITT Rayonier Inc.: Syracuse, NY, USA, 1983. Available online: https://www.osti.gov/biblio/5062478 (accessed on 2 May 2019).

42. Dufresne, A. Nanocellulose: A new ageless bionanomaterial. *Mater. Today* **2013**, *16*, 220–227. [CrossRef]

43. Nechyporchuk, O.; Belgacem, M.N.; Bras, J. Production of cellulose nanofibrils: A review of recent advances. *Ind. Crops Prod.* **2016**, *93*, 2–25. [CrossRef]

44. Lavoine, N.; Desloges, I.; Dufresne, A.; Bras, J. Microfibrillated cellulose—Its barrier properties and applications in cellulosic materials: A review. *Carbohydr. Polym.* **2012**, *90*, 735–764. [CrossRef]

45. Klemm, D.; Kramer, F.; Moritz, S.; Lindström, T.; Ankerfors, M.; Gray, D.; Dorris, A. Nanocelluloses: A new family of nature-based materials. *Angew. Chem. Int. Ed.* **2011**, *50*, 5438–5466. [CrossRef] [PubMed]

46. Brodin, F.W.; Gregersen, Ø.W.; Syverud, K. Cellulose nanofibrils: Challenges and possibilities as a paper additive or coating material—A review. *Nord. Pulp Pap. Res. J.* **2014**, *29*, 156–166. [CrossRef]

47. Oksman, K.; Aitomäki, Y.; Mathew, A.P.; Siqueira, G.; Zhou, Q.; Butylina, S.; Tanpichai, S.; Zhou, X.; Hooshmand, S. Review of the recent developments in cellulose nanocomposite processing. *Compos. Part A Appl. Sci. Manuf.* **2016**, *83*, 2–18. [CrossRef]

48. Hoeng, F.; Denneulin, A.; Bras, J. Use of nanocellulose in printed electronics: A review. *Nanoscale* **2016**, *8*, 13131–13154. [CrossRef] [PubMed]

49. Jorfi, M.; Foster, E.J. Recent advances in nanocellulose for biomedical applications. *J. Appl. Polym. Sci.* **2015**, *132*, 1–19. [CrossRef]

50. Venturi, D.; Grupkovic, D.; Sisti, L.; Baschetti, M.G. Effect of humidity and nanocellulose content on Polyvinylamine-nanocellulose hybrid membranes for CO_2 capture. *J. Membr. Sci.* **2018**, *548*, 263–274. [CrossRef]

51. Ansaloni, L.; Salas-Gay, J.; Ligi, S.; Baschetti, M.G. Nanocellulose-based membranes for CO_2 capture. *J. Membr. Sci.* **2017**, *522*, 216–225. [CrossRef]

52. Venturi, D.; Ansaloni, L.; Baschetti, M.G. Nanocellulose based facilitated transport membranes for CO_2 separation. *Chem. Eng. Trans.* **2016**, *47*, 349–354. [CrossRef]

53. Torstensen, J.; Helberg, R.M.L.; Deng, L.; Gregersen, Ø.W.; Syverud, K. PVA/nanocellulose nanocomposite membranes for CO_2 separation from flue gas. *Int. J. Greenh. Gas Control* **2019**, *81*, 93–102. [CrossRef]

54. Pääkko, M.; Ankerfors, M.; Kosonen, H.; Nykänen, A.; Ahola, S.; Österberg, M.; Ruokolainen, J.; Laine, J.; Larsson, P.T.; Ikkala, O.; et al. Enzymatic hydrolysis combined with mechanical shearing and high-pressure homogenization for nanoscale cellulose fibrils and strong gels. *Biomacromolecules* **2007**, *8*, 1934–1941. [CrossRef]

55. Spence, K.L.; Venditti, R.A.; Rojas, O.J.; Habibi, Y.; Pawlak, J.J. A comparative study of energy consumption and physical properties of microfibrillated cellulose produced by different processing methods. *Cellulose* **2011**, *18*, 1097–1111. [CrossRef]

56. Jonoobi, M.; Mathew, A.P.; Oksman, K. Producing low-cost cellulose nanofiber from sludge as new source of raw materials. *Ind. Crops Prod.* **2012**, *40*, 232–238. [CrossRef]

57. Josset, S.; Orsolini, P.; Siqueira, G.; Tejado, A.; Tingaut, P.; Zimmermann, T. Energy consumption of the nanofibrillation of bleached pulp, wheat straw and recycled newspaper through a grinding process. *Nord. Pulp Pap. Res. J.* **2014**, *29*, 167–175. [CrossRef]

58. Wågberg, L.; Decher, G.; Norgren, M.; Lindström, T.; Ankerfors, M.; Axnäs, K. The build-up of polyelectrolyte multilayers of microfibrillated cellulose and cationic polyelectrolytes. *Langmuir* **2008**, *24*, 784–795. [CrossRef] [PubMed]

59. Naderi, A.; Lindström, T.; Sundström, J. Repeated homogenization, a route for decreasing the energy consumption in the manufacturing process of carboxymethylated nanofibrillated cellulose? *Cellulose* **2015**, *22*, 1147–1157. [CrossRef]

60. Missoum, K.; Belgacem, M.N.; Bras, J. Nanofibrillated cellulose surface modification: A review. *Materials* **2013**, *6*, 1745–1766. [CrossRef] [PubMed]

61. Rol, F.; Belgacem, M.N.; Gandini, A.; Bras, J. Recent advances in surface-modified cellulose nanofibrils. *Prog. Polym. Sci.* **2019**, *88*, 241–264. [CrossRef]

62. Gebald, C.; Wurzbacher, J.A.; Tingaut, P.; Zimmermann, T.; Steinfeld, A. Amine-Based Nanofibrillated Cellulose As Adsorbent for CO_2 Capture from Air. *Environ. Sci. Technol.* **2011**, *45*, 9101–9108. [CrossRef] [PubMed]

63. Piccinini, E.; Giacinti Baschetti, M.; Sarti, G. Use of an automated spring balance for the simultaneous measurement of sorption and swelling in polymeric films. *J. Memb. Sci.* **2004**, *234*, 95–100. [CrossRef]

64. Crank, J. *The Mathematics of Diffusion*; Oxford University Press: Oxford, UK, 1975; p. 414.

65. Catalano, J.; Myezwa, T.; De Angelis, M.G.; Baschetti, M.G.; Sarti, G.C. The effect of relative humidity on the gas permeability and swelling in PFSI membranes. *Int. J. Hydrog. Energy* **2012**, *37*, 6308–6316. [CrossRef]

66. Kumar, S.; Rai, S.B. Spectroscopic studies of L-arginine molecule. *Indian J. Pure Appl. Phys.* **2010**, *48*, 251–255.

67. Davis, E.M.; Elabd, Y.A. Water Clustering in Glassy Polymers. *J. Phys. Chem. B* **2013**, *117*, 10629–10640. [CrossRef] [PubMed]

68. Park, G.S. Transport Principles—Solution, Diffusion and Permeation in Polymer Membranes. In *Synthetic Membranes: Science, Engineering and Applications*; Bungay, P.M., Lonsdale, H.K., de Pinho, M.N., Eds.; Springer: Dordrecht, The Netherlands, 1986; pp. 57–107.

69. Holz, M.; Heil, S.R.; Sacco, A. Temperature-dependent self-diffusion coefficients of water and six selected molecular liquids for calibration in accurate 1H NMR PFG measurements. *Phys. Chem. Chem. Phys.* **2000**, *2*, 4740–4742. [CrossRef]

70. Gouanvé, F.; Marais, S.; Bessadok, A.; Langevin, D.; Morvan, C.; Métayer, M. Study of water sorption in modified flax fibers. *J. Appl. Polym. Sci.* **2006**, *101*, 4281–4289. [CrossRef]
71. Minelli, M.; Baschetti, M.G.; Doghieri, F.; Ankerfors, M.; Lindström, T.; Siró, I.; Plackett, D. Investigation of mass transport properties of microfibrillated cellulose (MFC) films. *J. Membr. Sci.* **2010**, *358*, 67–75. [CrossRef]

 *nanomaterials*

Article

Composite Membranes Derived from Cellulose and Lignin Sulfonate for Selective Separations and Antifouling Aspects

Andrew Colburn [1], Ronald J. Vogler [1], Aum Patel [1], Mariah Bezold [1], John Craven [1], Chunqing Liu [2] and Dibakar Bhattacharyya [1,*]

[1] Department of Chemical and Materials Engineering, University of Kentucky, Lexington, KY 40506, USA; andrew.colburn@uky.edu (A.C.); voglerronald@uky.edu (R.J.V.); aum.patel@uky.edu (A.P.); mariah.bezold@uky.edu (M.B.); johndcraven@gmail.com (J.C.)

[2] R&D Department, Honeywell UOP, Des Plaines, IL 60016, USA; Chunqing.Liu@Honeywell.com

* Correspondence: db@uky.edu; Tel.: +(859)-257-2794

Received: 24 April 2019; Accepted: 4 June 2019; Published: 7 June 2019

Abstract: Cellulose-based membrane materials allow for separations in both aqueous solutions and organic solvents. The addition of nanocomposites into cellulose structure is facilitated through steric interaction and strong hydrogen bonding with the hydroxy groups present within cellulose. An ionic liquid, 1-ethyl-3-methylimidazolium acetate, was used as a solvent for microcrystalline cellulose to incorporate graphene oxide quantum dots into cellulose membranes. In this work, other composite materials such as, iron oxide nanoparticles, polyacrylic acid, and lignin sulfonate have all been uniformly incorporated into cellulose membranes utilizing ionic liquid cosolvents. Integration of iron into cellulose membranes resulted in high selectivity (>99%) of neutral red and methylene blue model dyes separation over salts with a high permeability of 17 LMH/bar. With non-aqueous (alcohol) solvent, iron–cellulose composite membranes become less selective and more permeable, suggesting the interaction of iron ions cellulose OH groups plays a major role in pore structure. Polyacrylic acid was integrated into cellulose membranes to add pH responsive behavior and capacity for metal ion capture. Calcium capture of 55 mg Ca^{2+}/g membrane was observed for PAA-cellulose membranes. Lignin sulfonate was also incorporated into cellulose membranes to add strong negative charge and a steric barrier to enhance antifouling behavior. Lignin sulfonate was also functionalized on the commercial DOW NF270 nanofiltration membranes via esterification of hydroxy groups with carboxyl group present on the membrane surface. Antifouling behavior was observed for both lignin-cellulose composite and commercial membranes functionalized with lignin. Up to 90% recovery of water flux after repeated cycles of fouling was observed for both types of lignin functionalized membranes while flux recovery of up to 60% was observed for unmodified membranes.

Keywords: nanocomposite; ionic liquid; selective separation; water application

---

1. Introduction

Cellulose is the most abundant biopolymer on the earth. The structural significance of cellulose toward life on Earth is profound, as it makes up most of the cell wall in plants, providing structural support. Cellulose within the cell wall of plants arranges itself in a mesoporous structure to sterically prevent enzymatic decomposition [1]. The robust characteristics of cellulose fibers have been utilized by humans for millennia. The use of cellulose fibers as a membrane-like material arguably began when humans first began making textiles out of cotton in the 6th millennium BC in modern day India and Pakistan [2]. Cellulose fibers remain an effective material for physical size-based separations of particulates such as algae clusters that act as carriers for cholera [3,4]. Beyond particle separation, the polymer network of cellulose materials has been investigated for the separation of smaller organic molecules. Transport of solutes through cellulose membranes has been of interest in the scientific community for many years. Dating back to the 1950s, hindered diffusion of small organic molecules in aqueous solution through cellulose materials such as cellophane and sausage casings were studied [5]. Cellulose-derived polymers such as cellulose acetate have been widely used for membrane making, but the modification required to enhance solubility in commercial solvents reduces robustness of membrane during filtration of organic solvents [6].

The strong hydrogen bonding between cellulose chains poses challenges for dissolution and processing of cellulosic material. New solvent developments have enabled synthesis of cellulose membranes without modification of the chemical structure of cellulose. Regeneration of cellulose with solvents such as N-methylmorpholine-N-oxide or basic conditions is being implemented to create selective membranes [7–9]. Ionic liquids are being investigated as a new solvent for regeneration of cellulose for membrane synthesis [10,11]. Membranes utilizing ionic liquid as a solvent have been shown to perform in the ultrafiltration or nanofiltration regimes. This same ionic liquid approach was used to spin cellulose hollow fibers [12]. Cellulose membranes prepared using ionic liquid have been shown to be highly selective for organic dyes, rejecting 94% of Bromothymol Blue [13]. Cellulose membranes have been shown to be effective at separating negatively charged oil emulsions from water while maintaining minimum fouling [14].

Composite materials have been integrated into polymer membranes to enhance properties and performance. Cellulose is a particularly interesting polymer for the integration of composite materials through hydrogen bonding via the hydroxy groups, which aid steric entanglement in nanocomposite retention. This allows for retention of composite materials that strongly interact with hydroxy groups such as layered double hydroxide [15]. Incorporation of graphene quantum dots into cellulose via ionic liquid was demonstrated to add negative charge and improve the selectivity of model dyes [16].

Beyond modifying membrane pore structure, the development of composite and blended materials brings additional properties that improve membrane performance and expand membrane application. Polyacids, particularly polyacrylic acid (PAA) have been incorporated into membrane supports to allow for strong negative charge, metal capture, and pH responsive behavior. PAA has been crosslinked within a PAA membrane pore to functionalize the pore with carboxyl groups for metal capture and in-situ nanoparticle functionalization [17]. PAA has also been grafted onto cellulose nanofiber mats for use in the capture of heavy metals [18]. Lignin, a complex plant-derived polymer, is another material that has capacity for heavy metal capture [19]. Due to the abundance of hydrophilic groups, lignin can be incorporated as a composite material with other polymers. Composite SPEEK/lignin membranes have been demonstrated to create a tighter pore structure than conventional Nafion membranes while allowing for enhanced proton transport [20].

The objective in this work was to further expand on our previous research studying cellulose graphene oxide quantum dots (GQD) membranes into other composite materials to further improve membrane performance and demonstrate flexibility of this technique for membrane development. Iron (III), polyacrylic acid, and lignin sulfonate were all investigated as composite materials for integration within the cellulose membrane domain. Membrane permeability and selectivity was studied for each composite type. Other useful properties unique to each composite material were observed such as metal capture and antifouling properties. Lignin sulfonate was also functionalized onto the surface of commercial NF membrane (NF270) to demonstrate antifouling behavior of the functionalized membrane surface through the creation of strong acid sulfonate groups on the surface.

2. Materials and Methods

2.1. Materials

1-ethyl-3-methylimidazolium acetate (EMIMAc, HPLC grade) was purchased from Sigma Aldrich (St. Louis, MO, USA. Avicel® PH-101 microcrystalline cellulose (50 μm, cotton linter source) was purchased from Sigma Aldrich. Nonwoven polyester backing material from Solecta Membranes was used as a support for membrane formation. Blue dextran (MW: 5000 Da; 10,000 Da) were purchased from Sigma Aldrich for use in membrane pore size characterization. Solutes used in selectivity studies can be seen in Table 1. Methylene Blue and Neutral Red (Sigma Aldrich) were used as model dies to study rejection of molecules <1000 Da. A model dimer (2-(2-Methoxyphenoxy)-1-(4-methoxyphenyl)ethanol) was provided by Dr. Mark Crocker's lab in the Center for Applied Energy Research. Ferric chloride (Fisher Scientific, Hanover Park, IL, USA) was used as an iron (III) source in composite membrane synthesis. Lignosulfonic acid sodium salt was purchased from Beantown Chemical LLC, Hudson, NH, USA. as a lignin sulfonate source. Humic acid (technical grade) and bovine serum albumin were purchased from Sigma Aldrich for antifouling study. Na2SO4 (1000 mg/L Fisher Scientific) was used to characterized nanofiltration membrane performance. The country of origin for all membranes and chemicals use was the United States of America.

Table 1. Model dyes tested for rejection.

Model Solute	Molecular Wt. (Da)	Structure
β-O-4 Model Dimer	282	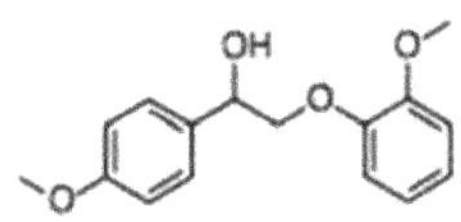
Neutral Red	289	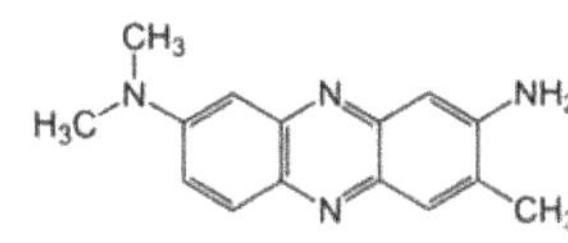
Methylene Blue	320	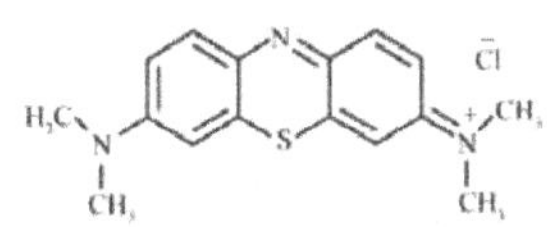

2.2. Cellulose Composite Membranes

Three types of Cellulose composite membranes were studied: cellulose iron, cellulose PAA, and cellulose lignin sulfonate composite membranes. A summary of the composition of the various membranes can be seen in Table 2. Control membranes of 10 wt% cellulose were also studied. All membranes were created using 1-ethyl-3-methylimidazolium as a solvent. The desired amount of composite material was dispersed into the ionic liquid at 80 °C for one hour. This is to ensure full dispersion of the composite material in the ionic liquid before dissolution of cellulose increases the casting solution viscosity. After composite material dispersion, 5–10 wt% cellulose was added into the casting solution and physically mixed in then dissolved at 80 °C for 8 to 24 h until the cellulose was completely dissolved.

Membranes were cast on nonwoven fiber backing. Polyester support material was affixed to a glass plate using tape. The casting solution was poured directly onto the backing at 80 °C and cast directly onto the polyester backing using a doctor blade set to 150 μm. The polyester backing was then submerged in a water gelation bath for 10 min to allow time for membrane formation. The resulting membrane was stored in DI water at a temperature of 4 °C until use.

Table 2. Composite membranes studied with compositions and relevant properties.

Composite Material	Wt% Composite	Wt% Cellulose	Casting Solution Viscosity (Pa*s)	Water Permeability (LMH/bar) (pH=7)	Rejection (%) 5kDa Blue Dextran
Iron	4	5	6.8	17.4	>99
PAA	2	5	44	267	44
Lignin	5	10	96	17.5	59
Cellulose only	0	10	22.8	9.6	75

2.3. Zeta Potential Characterization

Zeta potential of cellulose and GQD cellulose membranes was measured by an Anton Paar Surpass 1 Electrokinetic Analyzer, Ashland, VA, USA. The adjustable gap cell was used with a 100 μm gap and 0.01 M KCl electrolyte solution. Acid titration was done with 0.01 M HCl. A 400 mBar pressure difference was used for all measurements.

2.4. Contact Angle Characterization

The contact angle for deionized ultrafiltered water was measured using the Kruss DSA 100, Matthews, NC, USA. Captive bubble method was used to determine contact angle do to water absorption in the cellulose membranes and to prevent deformation of surface structure during drying. At least 3 spots per membrane sample were analyzed to correct for any variance in surface morphology.

2.5. Membrane Performance

Membrane performance was characterized by using the Sterlitech HP4750 stirred cell to perform convective studies. Water permeability was determined for each membrane by measuring the volumetric flux of deionized ultrafiltered (DIUF) water at 1.4, 2.76, and 4.14 bar respectively. Methylene blue (5 mg/L) and neutral red (5 mg/L), as well as various molecular weights (5 kDa and 10 kD at concentrations of 100 mg/L) of Blue Dextran, were filtered through the membrane. The permeate was collected and dye concentration for the feed, permeate, and remaining retentate was analyzed using the VWR UV-6300PC Spectrophotometer.

For cellulose–lignin composite membranes, antifouling properties were analyzed by testing permeability of 100 mg/L humic acid solution at pH of 5.6 in a crossflow system.

2.6. Divalent Ion Capture by Cellulose-PAA Membranes

Ca^{2+} capture in cellulose PAA composite membranes was carried out following the procedure used by Islam et al. [17]. The cellulose PAA membrane was added to a Sterlitech HP4750 filtration cell and soaked in about 110 mL of DI water at a pH of 10. After soaking, about 15 mL of fresh DI water (pH » 4.5–5.5) was passed through the membrane and the pH of the permeate was verified to be 7 or higher. For the Ca^{2+} capture, an aqueous $CaCl_2 \cdot 2H_2O$ solution (»1.79 mM, pH = 6.5–7) was prepared with non-deoxygenated, DI water and a 10-mL sample of this solution was taken. To capture Ca^{2+}, about 200 mL of fresh solution was passed through the membrane in 50-mL increments using pressures mostly in the range of 0.28–0.62 bar. At the end of each increment, a 10-mL sample of the collected permeate was taken and the rest of the permeate was disposed of before continuing the filtration.

Ca^{2+} captured was quantified by inductively coupled plasma optical emission spectroscopy. Ca^{2+} captured within the membrane case measured and located using energy-dispersive X-ray spectroscopy.

2.7. Lignin Sulfonate Functionalized Nanofiltration Membrane

Functionalized membranes were created by placing a 40 cm^2 area of NF270 into a circular metal cell, an O-ring is typically found on these cells at the base to create a seal. A 10 wt% LS (lignin sulfonate) solution in water was poured over a 40 cm^2 area NF270 nanofiltration membrane. Sufficient volume of lignin sulfonate was applied to cover the entire membrane surface. The cell was then placed in an oven at 90 °C for approximately 2 h. After taking the functionalized membrane out of the cell, it was then thoroughly rinsed with DI water to remove residual LS that may not have bonded to the membrane. LS presence on the membrane surface can be visually confirmed by light brown tint on the membrane surface.

A crossflow apparatus allowed for testing the anti-fouling properties of both the functionalized and unfunctionalized membrane. The cross-flow apparatus was run at a flowrate of approximately 1.5 liters/min for both the equilibration stage, fouling stage, and tangential washing stage. Before any anti-fouling testing could be done, the membrane was precompacted at 10.4 bar with deionized water to equilibrate the membrane before the fouling agent. After this equilibration period, a bovine albumin serum (BSA) solution was run through the apparatus and volume of permeate measured. After 30 min into the fouling stage, the membrane surface was rinsed with deionized water (pH = 5.6) for 10 min. Na_2SO_4 (1000 mg/L Fisher Scientific, Hanover Park, IL, USA) rejection was also determined in the crossflow cell at 10.4 bar.

2.8. Bacteria Fouling Studies

R. palustris strain CGA009 (ATCC BAA-98) was purchased from ATCC (American Type Culture Collection, NY, NY, USA). Solid media cultures were isolated on tryptic soy broth agar plates. Liquid cultures were pre-grown in tryptic soy broth purchased from Hardy Diagnostics, Santa maria, CA, USA, which contains (g L^{-1}) casein peptone, 17; soy peptone, 3; NaCl, 5; K_2HPO_4, 2.5; Dextrose, 2.5. Pre-grown liquid cultures were concentrated by centrifugation at 2500 rpm for 5 mins and washed 3 times with minimal media to use as an inoculant.

R. palustris adhered to membranes were grown using a modified minimal media [21] that contained (g L−1) Na_2HPO_4, 6.8; KH_2PO_4, 2.9; NaCl, 1.3; $MgSO_4$ $7H_2O$, 0.4; $CaCl_2$ $2H_2O$, 0.075; thiamine hydrochloride 0.001. Trace elements were provided by adding 10 mL L^{-1} of a solution containing (g L^{-1}) $FeCl_3$ $6H_2O$, 1.66; $ZnCl_2$, 0.17; $MnCl_2$, 0.06; $CoCl_2$ $6H_2O$, 0.06; $CuCl_2$ $2H_2O$, 0.04; $CaCl_2$ $2H_2O$, 0.73; and Na_2MoO_4 $2H_2O$, 0.06. sodium glutamate (3.5–7 mM), and acetate (70 mM) were utilized as primary nitrogen and carbon sources.

Solutions of minimal media were diluted 1:10 with phosphate buffer (pH~7.2) for inoculation on the membrane surface. Inoculation of cellulose membranes was carried out by convectively passing 15 mL of the diluted media through the membrane at 1.4 bar in a stirred membrane cell. After inoculation the membranes were removed from the cell and submerged in minimal growth media in the absence of light for 10 days to allow some time for bacterial growth. The overall goal was to simulate bacteria deposition and growth on the membrane surface over long-term operation.

Bacteria adhered membranes were chemically fixed [22] prior to critical point drying by dosing growth media containing an inoculated membrane with glutaraldehyde (50% from) to bring the solution to 2.5% glutaraldehyde, and left to sit for 2 h at 25 °C. The media slowly replaced with ethanol by removing media and adding ethanol to bring the ethanol concentration up to 25%, 50%, 75%, 85%, and 96%, leaving the solution to rest for 1 h between adding ethanol.

3. Results and Discussion

3.1. Summary of Membranes

A summary of composite membrane compositions and properties is given in Table 2. Permeability of all composite membrane was shown to be improved over unmodified cellulose membranes. Iron was the only composite material demonstrated to improve the selectivity of cellulose membranes for small molecules. There are many factors that impact membrane selectivity, only a few of which this work will address, but casting viscosity and wt% may be one property which can be better optimized to improve membrane performance. The focus on this work is to highlight the possibilities of incorporating composite materials into cellulose membranes and the unique benefits composite materials bring to membrane performance.

3.2. Iron Cellulose Composite Membranes

Prior work integrating graphene oxide quantum dots (GQDs) showed potential for suitably small nanomaterials to increase selectivity via improving the structure of the selective cellulose network by linking cellulose chains via hydrogen bonding. Nanocomposite membranes prepared with GQD were found to show over 80% rejection of methylene blue and greater than 95% rejection of 5000 kDa blue dextran while maintaining permeability over 12 LMH/bar. Due to difficulties in purification of GQD, it is desired to investigate other composite materials that can be incorporated at higher concentrations while maintaining controlled particle size.

Iron was readily incorporated as a composite material into the cellulose membrane domain, as $FeCl_3$ is highly dispersible in ionic liquids. Iron is well known to interact strongly with cellulose and to bind to cellulose chains. This interaction along with steric effects ensure iron is retained within the membrane structure after phase inversion. A clear sign of the presence of iron within the membrane can been seen by the orange color the iron brings to the normally translucent cellulose membrane. This can be seen in Figure 1. This effect has also been observed in our prior studies with graphene oxide quantum dots as nanocomposites.

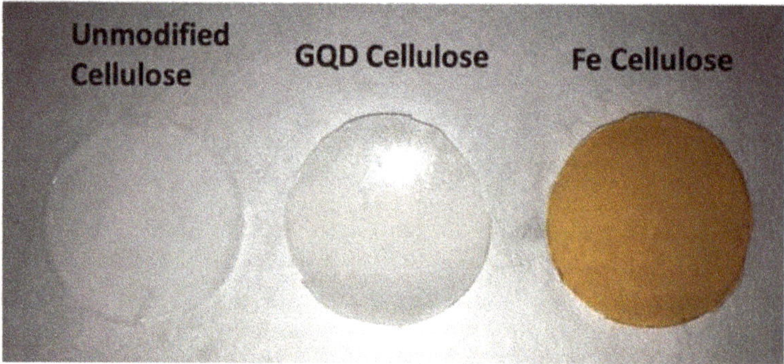

Figure 1. Unmodified cellulose, graphene oxide quantum dots (GQD) cellulose, and iron cellulose composite membranes.

While uses of iron in composite materials and membrane platforms is well known, the main interest in this work was to understand if iron interaction with cellulose in the membrane effects selectivity behavior of the membrane. Previous study of cellulose composite membranes has suggested that selectivity behavior is largely due to a dense amorphous polymer layer that comprises the top 100–200 nm of the membrane. To better understand how iron and cellulose might be interact in the amorphous selective layer, the pressure dependent flux behavior of the membrane was studied in water and IPA solvents. As seen in Figure 2, water and isopropanol permeability behavior is unique when compared to expected behavior for cellulose based membranes. Water flux plateaued off at higher pressures, as previously observed in our studies of GQD cellulose composite membranes. Permeability of the iron cellulose membranes was within standard deviation of previously studied cellulose membranes, despite the iron composite casting solution having half the concentration of cellulose, as compared to previously studied cellulose membranes. Ordinarily reduced concentration of cellulose in the casting solution results in higher permeability. It is observed that the 4 wt.% of iron (III) chloride in the casting solution contributes to the development of a dense layer at lower cellulose concentrations. It is important to note that the 5 mg/L neutral red solution was permeated through the membrane after IPA passage, demonstrating that the flux response is reversible with solvent exchange.

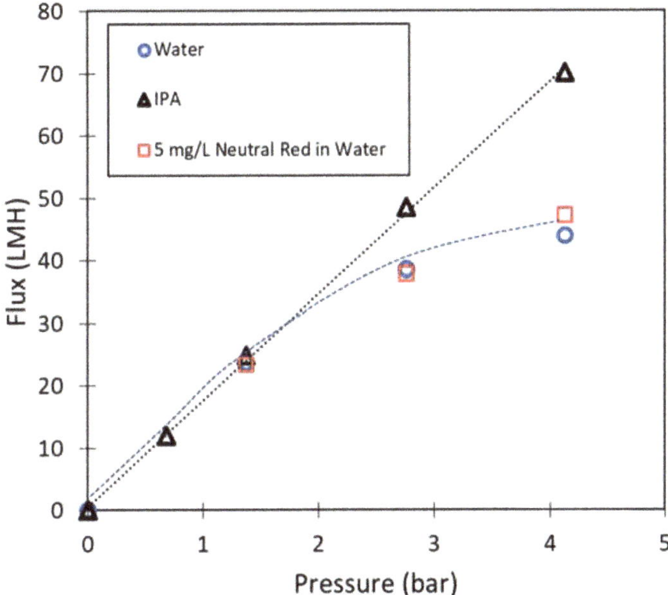

Figure 2. Flux (LMH, liter/m^2-bar) vs. pressure (bar) behavior for iron cellulose composite membranes in water, isopropanol (IPA), and neutral red in water.

Further investigation of pressure dependent isopropanol flux gave unexpected results. Despite the viscosity of IPA being roughly double that of water, the permeability remains the same. Isopropanol permeability had previously been studied in cellulose membranes as seen in Figure 3 to observe the effect of the polar solvent on membrane stability. When corrected for viscosity, isopropanol and water flux behavior follow the same linear trend. This strongly suggests the membrane surface does not swell when in contact with isopropanol. IPA flux was higher than what would be expected when observing water permeability. This suggests that the membrane becomes more permeable when exposed to isopropanol.

Flux behavior was also found to be reversible as solvent was varied between water and isopropanol. No significant concentrations of iron ions were found to leach out of the membrane beyond the initial formation of the membrane via phase inversion. Considering lack of evidence of significant leaching and reversible flux response during solvent exchange, iron cellulose composite membranes can be assumed to be reasonably stable in neutral or basic conditions. Water was kept above pH 5 for all iron cellulose composite membrane experiments. At acidic pH complete ionization of iron oxide may result in increased leaching of iron from the membrane domain.

Interestingly, water permeability was decreased at solvent mixtures of 25:75 and 50:50 isopropanol:water as compared to either pure water or pure isopropanol. The permeability at different solvent concentrations can be seen in Figure 4. Strong water interaction within the cellulose domain may provide a barrier for isopropanol diffusion into the membrane domain. Mao et al. has observed that flux through cellulose membranes declined as isopropanol concentration increased during pervaporation operation [9]. Membrane permeability increased for pure isopropanol solvent, as iron is unable to be ionized after the 100% isopropanol solution removes residual water in the membrane. Absence of Fe^{3+} ions decreases interaction between cellulose chains, which may be responsible for the higher permeability in pure solvent conditions. When the membrane is rehydrated, more ionization of iron particles to Fe^{3+} occurs and the selective layer becomes denser. This behavior could be of great interest for applications of membrane cleaning or desorption of contaminants from the membrane surface.

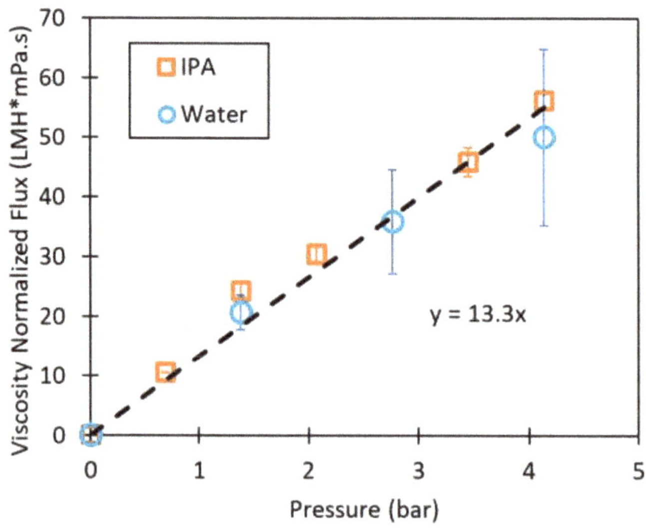

Figure 3. Viscosity corrected flux vs. pressure for unmodified cellulose membrane (10 wt. %). Temp 25 °C.

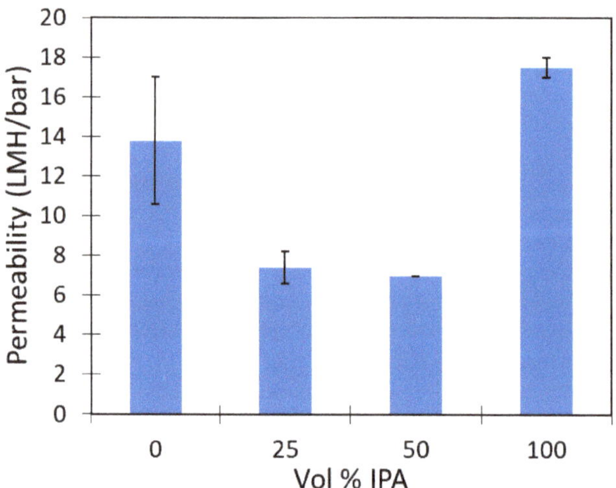

Figure 4. Volumetric permeability of total solvent mixture as volume % of isopropanol is varied in iron cellulose composite membranes. Remaining volume % water.

Neutral red (~289 Da) and methylene blue (~320 Da) were completely rejected (>99%) during filtration through the membrane using DI water as a solvent. As seen in Figure 5 rejection decreased in isopropanol which is to be expected due as hydrophilic interaction decreases in isopropanol. The increase of membrane permeability suggests the dense selective layer becomes more permeable.

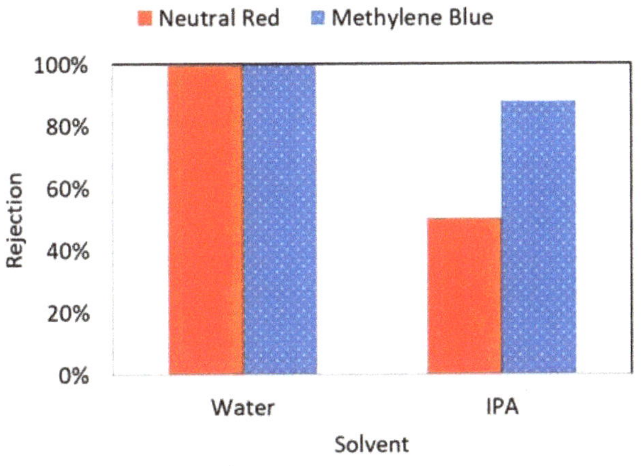

Figure 5. Dye rejection in iron cellulose composite membrane in water and isopropanol solvent.

Rejection studies with model dyes also suggests other factors contribute to solute selectivity other than size exclusion. Selectivity vs. molecular weight for small model molecules is show in Figure 6. Rejection of β-O-4 Model Dimer was only 10% despite the MW only being 7 Da less than neutral red. The disparity in rejection can be attributed to interaction among the hydrophilic functional groups. The positive dipoles of the amine groups in the dyes interacts more strongly with negative dipoles of hydroxy groups in cellulose reducing rate of diffusion of the dyes through the membrane. Carboxyl groups in the model dimer do not react as strongly. Rotational freedom in the model dimer may also allow for the dimer to change confirmation as it moves through the membrane, thus increasing diffusion rate. Ring structures in the model dyes prevent rotation within dyes as they move into the membrane domain. Interaction among functional groups and molecular structure must be considered when evaluating possible application of nanofiltration for small molecule separation.

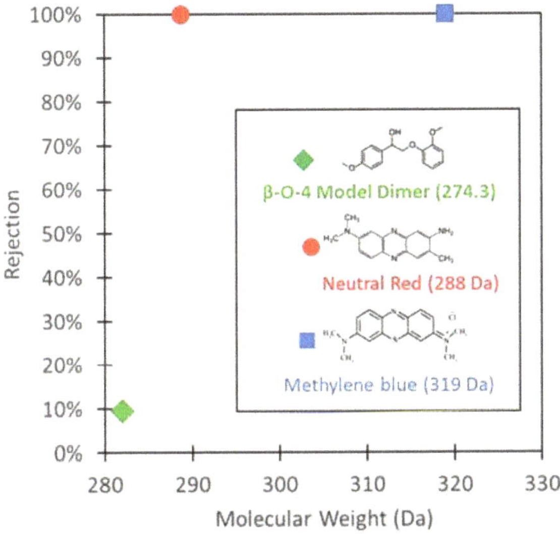

Figure 6. Rejection of model dyes and molecules in iron cellulose composite membranes.

3.3. Polyacrylic Acid Cellulose Composite Membranes

Polyacrylic acid (PAA) has many negatively charged carboxyl groups which can be utilized for pH responsive behavior, metal capture, and rejection of negatively charged ions. PAA disperses fully in the ionic liquid solvent allowing even mixing with cellulose. Entanglement with cellulose chains and hydrogen bonding with cellulose allow for the retention of PAA after phase inversion. The pKa of carboxyl groups was useful in confirming its presence of PAA at the surface of the PAA cellulose composite membranes. Zeta potential analysis (Figure 7) clearly shows that incorporating PAA into cellulose membranes results in a greater magnitude of negative surface charge that corresponds to a pKa shift at pH 3–5, as expected for carboxyl groups. This behavior has been demonstrated in our previous studies with PAA functionalized PVDF (polyvinylidene fluoride) microfiltration membranes. Due to the dissolution of PAA and cellulose together in ionic liquid, it is hypothesized that PAA was also integrated through the depth of the membrane.

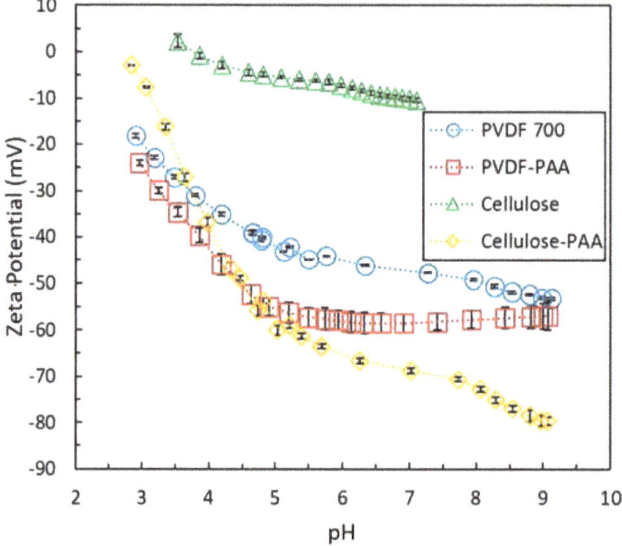

Figure 7. Zeta potential of cellulose (10 wt% cellulose in casting solution)[1], and cellulose- polyacrylic acid (PAA) membranes in the pH range of 3–9. PVDF 700 membrane obtained from Solecta Membrane and PVDF-PAA (weight gain of 7.28% with functionalization) were synthesized for this study following the procedure established in Islam et al. as a control system to demonstrate the impact of PAA functionalization on membrane surface charge [17].

Further confirmation of PAA in the membrane was necessary to confirm presence beyond the surface. Pressure dependent flux of PAA cellulose membranes were studied at below and above the pKa of PAA. As observed in other PAA functionalized membranes, swelling should occur as carboxyl groups are charged when pH increases above 3. Figure 8 shows the pH responsive behavior of the functionalized membrane. The four-fold decrease in flux when transitioning to pH 7 from pH 3 strongly suggests presence of PAA throughout the entire selective layer of the composite membrane. At high pH the swollen PAA creates a selective layer capable of rejecting 44% of 5kDa blue dextran, while at low pH the PAA collapses, opening the membrane pores.

PAA has been utilized for capture of metals due to the ion exchange capacity of the vast network of carboxylic groups. Ion exchange capacity studied for this membrane using Ca^{2+} to better understand the quantity of PAA in the membrane and the accessibility of PAA to ions transporting through. Previous functionalized membrane platforms have not completely answered the question of whether the entirety of the hydrogel is available for ion exchange, or whether channeling occurs within the hydrogel domain. In this scenario PAA is entangled along with the cellulose composite membrane which should theoretically prevent channeling. Ca^{2+} adsorption is shown in Figure 9. This was not observed in cellulose membranes, likely due to the constrained environment in which the PAA is present. The cellulose-PAA membrane demonstrated a maximum Ca^{2+} capture of 0.27 mg, which was equal to 0.055 g Ca^{2+}/g membrane and 0.35 mol Ca^{2+}/mol carboxyl; this result was confirmed by a mass balance that indicated a 98.2% retention of the fed Ca^{2+} in either the permeate samples, the retentate, or the membrane as well as a difference of 6.6% in measured concentration between the 50 ppm calibration curve sample and the duplicate of this sample. Ca^{2+} capture was roughly 70% of the theoretical maximum and just over half of the 0.61 mol Ca^{2+}/mol carboxyl that was calculated for the spongy PVDF-PAA membrane reported in literature [23]. Ca^{2+} capture observed in spongy PVDF-PAA membranes exceeded the theoretical value due to counter ion condensation phenomena within the membrane.

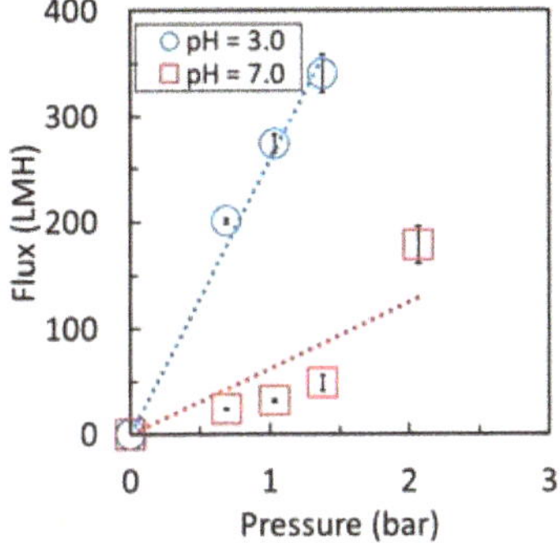

Figure 8. Volumetric flux of deionized water through of cellulose-PAA membrane at pH 3 and 7. Membrane surface area = 13.2 cm^2.

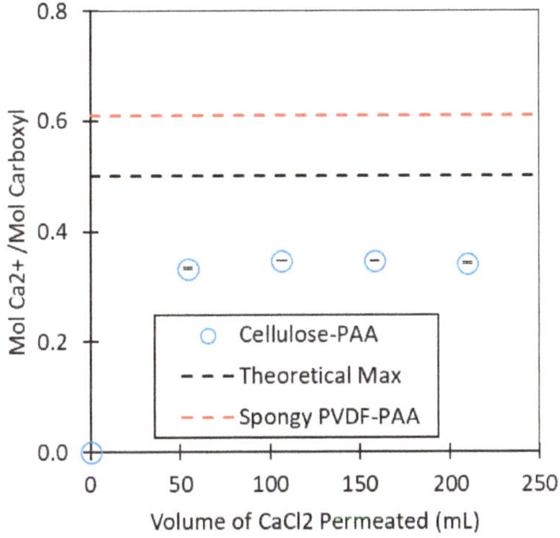

Figure 9. Total Ca^{2+} capture of a 13.2-cm^2 cellulose-PAA membrane during convective flow of $CaCl_2$ (overall flux = 89 LMH and average pressure of 50-mL increments = 0.72 bar) and of a PVDF-PAA membrane from literature after convective flow of $CaCl_2$ [23]. Standard deviation was determined via deviation of known samples after spiking with a separate known and thus represents analytical error during inductively coupled plasma optical emission spectroscopy (ICP-OES).

Electron dispersive x-ray spectroscopy of the PAA cellulose composite membrane was conducted to determine where metal ion capture was occurring within the membrane. The EDS mapping reveals that PAA cellulose membranes show even dispersion of divalent ions adsorbed throughout the membrane, while PAA functionalized PVDF membranes show divalent ion adsorption only toward the surface of the membrane. The EDS map (Figure 10) serves as further confirmation that PAA is evenly dispersed throughout the membrane.

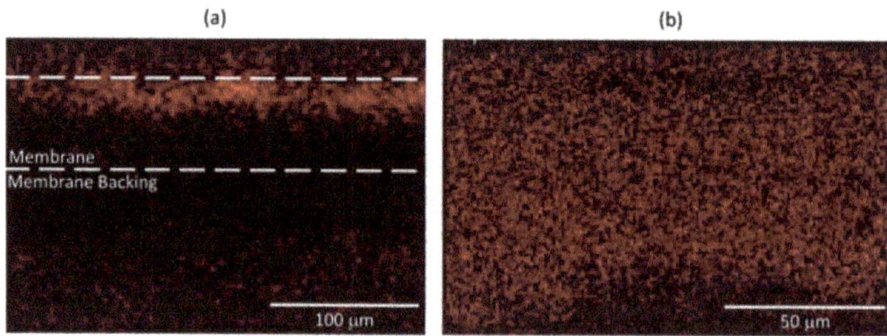

Figure 10. (**a**) Iron energy-dispersive X-ray spectroscopy (EDS) map of the cross section of a PVDF-PAA-Fe sample and (**b**) calcium EDS map of most of the cross section of a cellulose-PAA-Ca^{2+} sample.

3.4. Lignin Sulfonate Cellulose Composite Membranes

Lignin and cellulose are major constituents of woody plants and interact to create a robust structure that is resistant to decomposition from bacteria and fungi even after the plant's death. Lignin contains many hydrophilic groups, including phenols which give antibacterial properties [24,25]. Houtman et al. have determined through molecular simulation that hydrophilic groups allow for lignin to adsorb to cellulose microfibrils [26]. Lignin sulfonate, a byproduct of chemical paper pulping industry, is an inexpensive and commercially available source of lignin. The sulfonation process adds hydrophilicity and allows for easy dissolution in 1-ethyl-3-methylimidazolium acetate [27]. Therefore, we sought to use lignin sulfonate as a composite material for cellulose membrane creation. The primary objectives were to determine the effectiveness of the lignin composite membrane and probe antibacterial behavior.

Water permeability of the lignin cellulose membrane was shown to be roughly double that of the unmodified cellulose membrane (Figure 11). Likely hydrophobic regions of lignin sulfonate cause opening of the selective layer due to poor interaction with cellulose after phase inversion. The viscosity of the dope solutions was particularly high when lignin sulfonate was added as a composite, which may further effect demixing during phase inversion. The rejection of 5000 Da blue dextran was 59%, and this is 16% lower than unmodified cellulose. Neutral red was shown to absorb strongly in within the membrane, which indicates the potential of strong interaction with sulfonate groups within the membrane.

In order to characterize the fouling of a 10 wt% cellulose in IL membrane functionalized with 2 wt% lignin sulfonate, both a 10 wt% cellulose in IL membrane and a 10 wt% cellulose in IL membrane with 2 wt% lignin sulfonate (cellulose–lignin membrane) were placed in a cross-flow cell. 100 mg/L humic acid solution was passed through the cross-flow cell and the flux of both membranes was recorded over a total of 350 min. The black dotted lines indicate when both membranes were rinsed for 10 min with deionized water (pH of 5.6) to measure the effect of irreversible fouling. The normalized water flux of both the control membrane and the functionalized membrane are contained in Figure 12. The flux of both the control membrane and the flux of the membrane functionalized with lignin sulfonate decreased as more humic acid solution was passed. Error bars for each point indicate high water flux reproducibility. However, after each tangential rinse with deionized water, there is far less irreversible fouling of the functionalized membrane when compared to the control membrane. After only 10 min of rinsing, the functionalized membranes almost completely returned to the initial volumetric flux as recorded before fouling, while the control membrane shows only about 50% recovery of volumetric flux. Although the lignin sulfonate appears to have a negligible effect on the reversible fouling of the membrane, the lignin sulfonate does have a significant effect on reducing the prevalence of irreversible fouling of the functionalized membrane.

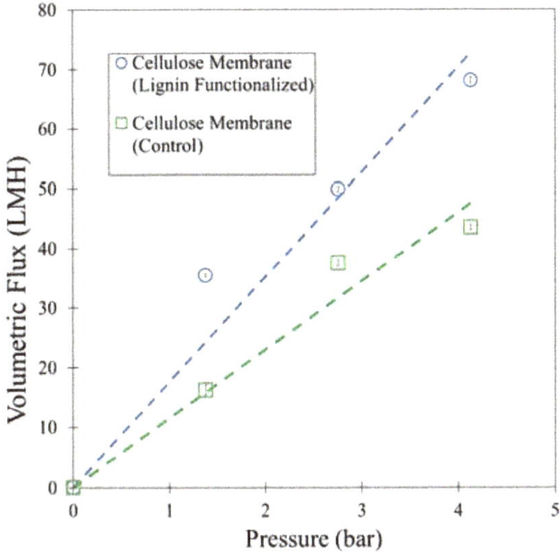

Figure 11. Pressure dependent volumetric water flux through cellulose membranes modified with lignin sulfonate and unmodified cellulose membranes (control).

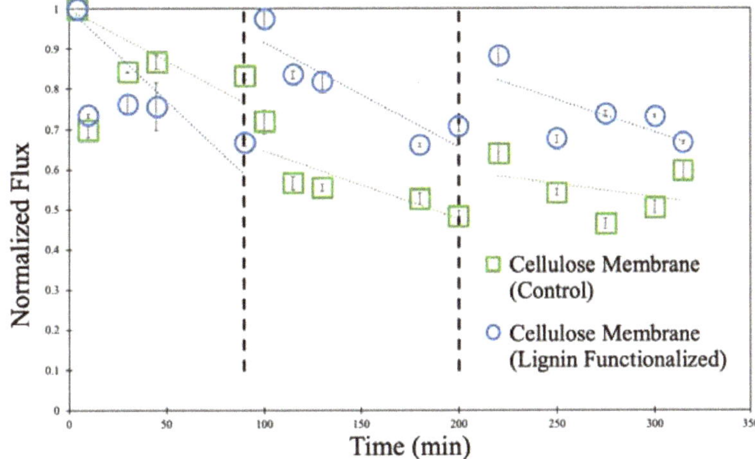

Figure 12. Normalized water flux of cellulose membrane functionalized with lignin sulfonate and of control cellulose membrane during filtration of 100 mg/L humic acid solution. Normalized flux = water flux with humic acid/pure water flux. pH = 5. Operating pressure = 10.4 bar. Vertical dashed lines indicate points during the experiment where tangential washing (1.5 L/min) with deionized ultrafiltered (DIUF) was performed to recover membrane flux. Trend lines (dotted) and flux error bars are shown in the figure.

Beyond natural organic matter and proteins, a key contributor to membrane fouling are biofilms formed by microorganisms that adsorb to the membrane surface. Since, it has been understood for many years that functional phenolic groups provide antimicrobial properties in lignin, the antimicrobial properties of lignin sulfonate modified cellulose membrane was studied [28]. Lignin sulfonate modified cellulose membranes were inoculated with bacteria by filtering a dilute solution of bacteria through the membrane. The bacterial were then given dilute amounts of nutrients and allowed to grow. Bacteria colonies were analyzed after fixation to qualitatively determine the rate of production of extracellular matrix. The SEM images of the membrane surface after bacteria growth can be seen in Figure 13.

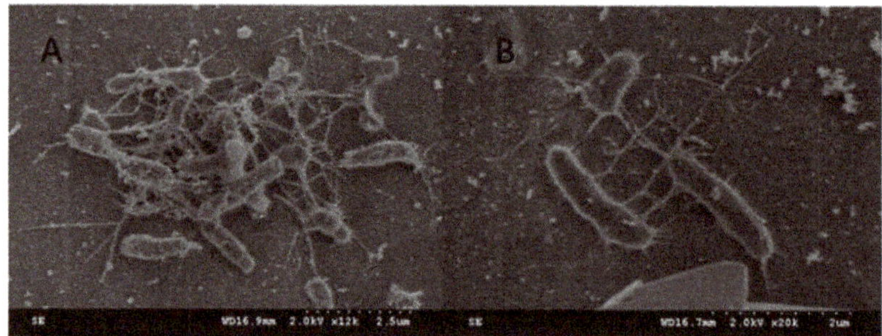

Figure 13. Bacteria growth on (**A**) unmodified cellulose membrane and (**B**) lignin sulfonate modified cellulose membrane.

3.5. Lignin Sulfonate Functionalized Commercial Nanofiltration Membrane

Lignin sulfonate can also be directly functionalized onto the surface of commercial nanofiltration membranes. Sulfonated lignin has shown potential antifouling properties when deposited onto the surface of thin film composite membranes. This study looked to use heat to esterify lignin to the surface of NF (nanofiltration) membranes. Membrane water permeability was shown to decreases slightly after functionalization (Figure 14), but flux decline was less than 10 wt%. This decline in flux was likely due to the surface functionalized layer adding resistance to flow through the membrane. Lignin has a bulky branching structure that could cause additional hydraulic resistance to flow.

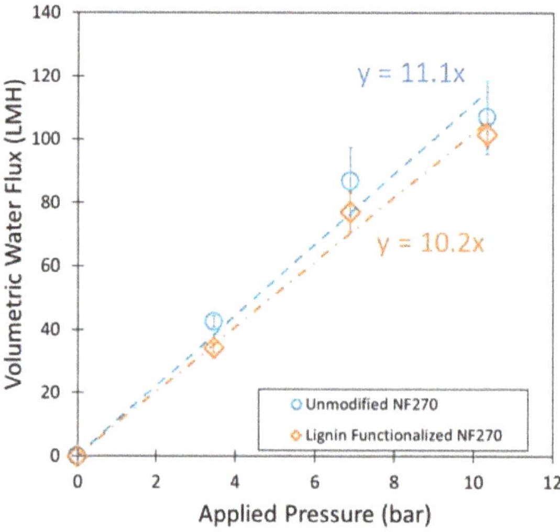

Figure 14. Pressure dependent water flux of unmodified NF270 and LS functionalized membrane.

Rejection of Na_2SO_4 (1000 mg/L solution) decreased to 97.3% from 98% after functionalization with lignin sulfonate, which is within experimental error for the small samples of membrane tested

(20 cm^2). Zeta potential data also suggests reduction in the number of carboxyl groups on the surface of the NF membrane (Figure 15).

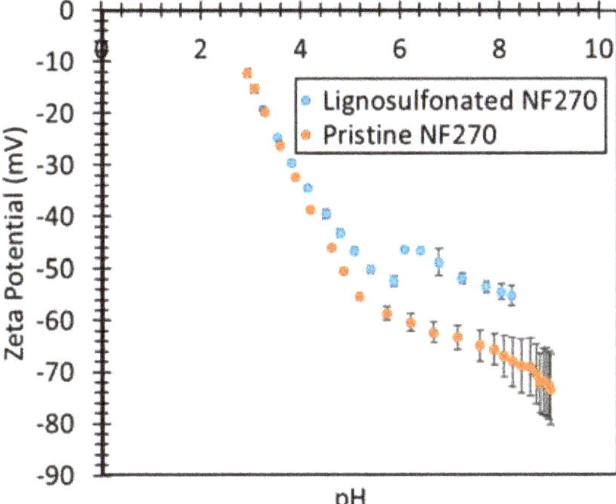

Figure 15. Zeta potential vs. pH for lignin functionalized and pristine NF270 membrane. 100 mg/L KCl used as an electrolyte.

Most excitingly, lignin sulfonate functionalized membranes show promise for use as an antifouling surface. BSA was used as a model foulant and passed through the membrane in cross flow operation. BSA fouling during filtration can be seen in Figure 16. While lignin sulfonate appears to have negligible impact on reversible fouling, irreversible fouling was shown to be far less prevalent after functionalization with lignin sulfonate. Functionalized NF270 membranes showed almost complete recovery of volumetric water flux after just 10 min of tangential flow rinsing with DI water while the unmodified membrane flux only recovered to 40% of the initial value after rinsing. Lignin functionalized NF270 membranes were shown to maintain 90% of the initial flux after the second rinse cycle.

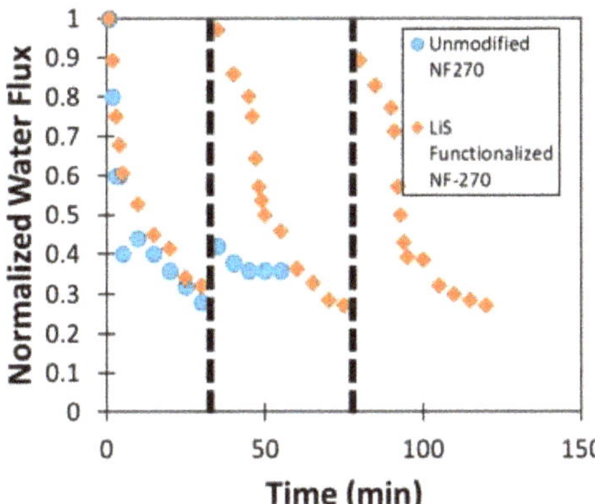

Figure 16. Normalized water flux of lignin sulfonate functionalized and unmodified NF270 during filtration of 100 mg/L BSA. Dotted lines indicate 10 min of tangential rinsing with deionized water (pH = 5.6).

4. Conclusions

This study has shown 1-ethyl-3-methyl imidazolium acetate ionic liquid can be utilized as a cosolvent to integrate iron, polyacrylic acid (PAA), or lignin sulfonate with cellulose membrane. Composite and blended materials were found to add unique properties such as pH responsive flux and antibacterial behavior. Performance of iron–cellulose composite membranes demonstrates that composite materials modify membrane structure and impact transport of solvent and solute through the membrane. Membrane structure was observed to become less selective in solvent conditions where affinity between iron and cellulose is reduced. Both steric entrapment and hydrogen bonding allow for PAA to be incorporated into the cellulose membrane domain for hardness ion capture applications. In the same manner, lignin sulfonate was incorporated covalently to reduce irreversible fouling on the membrane surface. This antifouling behavior was also observed when lignin sulfonate was functionalized onto the surface of the commercial NF270 membrane. Ultimately, as ionic liquids continue to be used as solvents for membrane synthesis, composite material should be strongly considered as means to add value or otherwise optimize membranes. Even inexpensive materials such as iron or sulfonated lignin have shown potential as composite materials, and impart little additional costs compared to the price of ionic liquid.

Author Contributions: All authors were involved in executing the work detailed in the manuscript. A.C., C.L., and D.B. were involved in the overall experiment execution and manuscript preparation, and the undergraduate students (R.J.V., A.P., and M.B.) were involved in cellulose PAA and cellulose–lignin sulfonate membranes synthesis. J.C. was involved in antimicrobial studies.

Funding: This research was funded by the NSF KY EPSCoR program (Grant number: 1355438) and by NIEHS-SRP program. Honeywell corporation funding provided partial support for lignin sulfonate functionalized membranes.

Conflicts of Interest: The authors declare no conflict of interest.

References

1. Adani, F.; Papa, G.; Schievano, A.; Cardinale, G.; D'Imporzano, G.; Tambone, F. Nanoscale structure of the cell wall protecting cellulose from enzyme attack. *Environ. Sci. Technol.* **2011**, *45*, 1107–1113. [CrossRef] [PubMed]
2. Moulherat, C.; Tengberg, M.; Haquet, J.-F.; Mille, B.T. First evidence of cotton at Neolithic Mehrgarh, Pakistan: Analysis of mineralized fibres from a copper bead. *J. Archaeol. Sci.* **2002**, *29*, 1393–1401. [CrossRef]
3. Colwell, R.R.; Huq, A.; Islam, M.S.; Aziz, K.M.A.; Yunus, M.; Khan, N.H.; Mahmud, A.; Sack, R.B.; Nair, G.B.; Chakraborty, J.; et al. Reduction of cholera in Bangladeshi villages by simple filtration. *Proc. Natl. Acad. Sci. USA* **2003**, *100*, 1051–1055. [CrossRef] [PubMed]
4. Huq, A.; Yunus, M.; Sohel, S.S.; Bhuiya, A.; Emch, M.; Luby, S.P.; Russek-Cohen, E.; Nair, G.B.; Sack, R.B.; Colwell, R.R. Simple sari cloth filtration of water is sustainable and continues to protect villagers from cholera in Matlab, Bangladesh. *MBio* **2010**, *1*, e00034-10. [CrossRef] [PubMed]
5. Renkin, E.M. Filtration, diffusion, and molecular sieving through porous cellulose membranes. *J. Gen. Physiol.* **1954**, *38*, 225–243. [CrossRef] [PubMed]
6. Lonsdale, H.K.; Merten, U.; Riley, R.L. Transport properties of cellulose acetate osmotic membranes. *J. Appl. Polym. Sci.* **1965**, *9*, 1341–1362. [CrossRef]
7. Yang, G.; Xiong, X.; Zhang, L. Microporous formation of blend membranes from cellulose/konjac glucomannan in NaOH/thiourea aqueous solution. *J. Membr. Sci.* **2002**, *201*, 161–173. [CrossRef]
8. Anokhina, T.S.; Yushkin, A.A.; Makarov, I.S.; Ignatenko, V.Y.; Kostyuk, A.V.; Antonov, S.V.; Volkov, A.V. Cellulose composite membranes for nanofiltration of aprotic solvents. *Pet. Chem.* **2016**, *56*, 1085–1092. [CrossRef]
9. Mao, Z.; Cao, Y.; Jie, X.; Kang, G.; Zhou, M.; Yuan, Q. Dehydration of isopropanol–water mixtures using a novel cellulose membrane prepared from cellulose/N-methylmorpholine-N-oxide/H_2O solution. *Sep. Purif. Technol.* **2010**, *72*, 28–33. [CrossRef]
10. Chen, H.Z.; Wang, N.; Liu, L.Y. Regenerated cellulose membrane prepared with ionic liquid 1-butyl-3-methylimidazolium chloride as solvent using wheat straw. *J. Chem. Technol. Biotechnol.* **2012**, *87*, 1634–1640. [CrossRef]

11. Lu, Y.; Liu, Y.X.; Yu, H.P.; Sun, Q.F. Preparation, characterization and biocompatibility of regenerated cellulose/PVA blend membranes in 1-allyl-3-methyimidazolium chloride. In *Materials Science and Engineering, Pts 1–2*; Zhu, G., Ed.; Trans Tech Publications: Zurich, Switzerland, 2011; Volume 179, pp. 179–180.

12. Lei, L.; Lindbråthen, A.; Sandru, M.; Gutierrez, M.; Zhang, X.; Hillestad, M.; He, X. Spinning cellulose hollow fibers using 1-ethyl-3-methylimidazolium acetate–dimethylsulfoxide co-solvent. *Polymers* **2018**, *10*, 972. [CrossRef] [PubMed]

13. Sukma, F.M.; Çulfaz-Emecen, P.Z. Cellulose membranes for organic solvent nanofiltration. *J. Membr. Sci.* **2018**, *545*, 329–336. [CrossRef]

14. Kim, D.; Livazovic, S.; Falca, G.; Nunes, S.P. Oil–water separation using membranes manufactured from cellulose/ionic liquid solutions. *ACS Sustain. Chem. Eng.* **2019**, *7*, 5649–5659. [CrossRef]

15. Iftekhar, S.; Srivastava, V.; Sillanpää, M. Synthesis and application of LDH intercalated cellulose nanocomposite for separation of rare earth elements (REEs). *Chem. Eng. J.* **2017**, *309*, 130–139. [CrossRef]

16. Colburn, A.; Wanninayake, N.; Kim, D.Y.; Bhattacharyya, D. Cellulose-graphene quantum dot composite membranes using ionic liquid. *J. Membr. Sci.* **2018**, *556*, 293–302. [CrossRef]

17. Islam, M.S.; Hernández, S.; Wan, H.; Ormsbee, L.; Bhattacharyya, D. Role of membrane pore polymerization conditions for pH responsive behavior, catalytic metal nanoparticle synthesis, and PCB degradation. *J. Membr. Sci.* **2018**, *555*, 348–361. [CrossRef] [PubMed]

18. Chitpong, N.; Husson, S.M. Nanofiber ion-exchange membranes for the rapid uptake and recovery of heavy metals from water. *Membranes* **2016**, *6*, 59. [CrossRef] [PubMed]

19. Ge, Y.; Li, Z.; Kong, Y.; Song, Q.; Wang, K. Heavy metal ions retention by bi-functionalized lignin: Synthesis, applications, and adsorption mechanisms. *J. Ind. Eng. Chem.* **2014**, *20*, 4429–4436. [CrossRef]

20. Ye, J.; Cheng, Y.; Sun, L.; Ding, M.; Wu, C.; Yuan, D.; Zhao, X.; Xiang, C.; Jia, C. A green SPEEK/lignin composite membrane with high ion selectivity for vanadium redox flow battery. *J. Membr. Sci.* **2019**, *572*, 110–118. [CrossRef]

21. Zhao, J.; Baba, T.; Mori, H.; Shimizu, K. Effect of *zwf* gene knockout on the metabolism of *Escherichia coli* grown on glucose or acetate. *Metab. Eng.* **2004**, *6*, 164–174. [CrossRef] [PubMed]

22. Chao, Y.; Zhang, T. Optimization of fixation methods for observation of bacterial cell morphology and surface ultrastructures by atomic force microscopy. *Appl. Microbiol. Biotechnol.* **2011**, *92*, 381–392. [CrossRef] [PubMed]

23. Davenport, D.M.; Gui, M.; Ormsbee, L.R.; Bhattacharyya, D. Development of PVDF membrane nanocomposites via various functionalization approaches for environmental applications. *Polymers* **2016**, *8*, 32. [CrossRef] [PubMed]

24. Vanholme, R.; Demedts, B.; Morreel, K.; Ralph, J.; Boerjan, W. Lignin biosynthesis and structure. *Plant Physiol.* **2010**, *153*, 895. [CrossRef] [PubMed]

25. García, A.; Spigno, G.; Labidi, J. Antioxidant and biocide behaviour of lignin fractions from apple tree pruning residues. *Ind. Crops Prod.* **2017**, *104*, 242–252. [CrossRef]

26. Houtman, C.J.; Atalla, R.H. Cellulose-lignin interactions (a computational study). *Plant Physiol.* **1995**, *107*, 977–984. [CrossRef] [PubMed]

27. Aro, T.; Fatehi, P. Production and Application of Lignosulfonates and sulfonated lignin. *Chem. Sustain. Chem.* **2017**, *10*, 1861–1877. [CrossRef] [PubMed]

28. Zemek, J.; Košíková, B.; Augustín, J.; Joniak, D. Antibiotic properties of lignin components. *Folia Microbiol.* **1979**, *24*, 483–486. [CrossRef]

 *nanomaterials*

Article

Electrospinning of Cellulose Nanocrystal-Filled Poly (Vinyl Alcohol) Solutions: Material Property Assessment

J. Elliott Sanders [1],*, Yousoo Han [1], Todd S. Rushing [2] and Douglas J. Gardner [1],*

[1] Advanced Structures and Composites Center (ASCC), School of Forest Resources (SFR), University of Maine (UMaine), 35 Flagstaff Rd., Orono, ME 04469, USA; yousoo.han@maine.edu

[2] Army Engineer Research and Development Center (ERDC), 3909 Halls Ferry Rd., Vicksburg, MS 39180, USA; todd.s.rushing@usace.army.mil

* Correspondence: jsande80@maine.edu (J.E.S.); douglasg@maine.edu (D.J.G.); Tel.: +1-207-581-2402 (J.E.S.); +1-207-581-2123 (D.J.G.)

Received: 26 April 2019; Accepted: 23 May 2019; Published: 27 May 2019

Abstract: Poly (vinyl alcohol) (PVA) and cellulose nanocrystals (CNC) random composite mats were prepared using the electrospinning method. PVA/CNC mats were reinforced with weight concentrations of 0, 20 and 50% CNC (*w/w*) relative to PVA. Scanning electron microscopy was used to measure the fiber diameter, which ranged from 377 to 416 nm. Thermogravimetric analysis (TGA) confirmed the presence of CNC fibers in the mat fibers which were not visible by scanning electron microscope (SEM). Mechanical testing was conducted using ASTM D 638 on each sample group at 10 mm min^{-1}. Neat PVA and PVA/CNC mats were heat treated at 170 °C for 2h hours, and the morphological structure was maintained with some fiber diameter reduction. Mechanical property results after heat treatment showed a decrease in tensile strength, an increase in tensile stiffness and a decrease in strain to yield (%). This effect was attributable to enhanced diffusion bonding of the mat fiber intersections. The CNC fibers also increased mat stiffness, and reduced strain to yield in non-treated mats. The use of CNCs show potential for compounding into bulk polymer composites as a reinforcement filler, and also show promise for chemical crosslinking attributable to the –OH groups on both the PVA, in addition to esterification of the vinyl group, and CNC.

Keywords: poly (vinyl alcohol); cellulose nanocrystals; electrospinning; polymer nanocomposites; tensile properties; scanning electron microscopy; rheology; thermogravimetric analysis

---

1. Introduction

Since the late 20th century, electrospinning has become a practically useful method for producing value added materials from dissolvable plastics, and more recently a method for producing nanocomposite materials [1]. Solution parameters influencing the electrospinning process include polymer concentration, solvent evaporation time, charge carrying capacity (conductivity), pH, average molecular weight (M_w) and the degree of hydrolysis (DH) of the polymer. A universal theory was also outlined which details how entanglement molecular weight (M_e) is useful for determining concentration levels capable of forming continuous electrospun filaments a priori. Shenoy et al. [2] utilized an equation which is used to calculate the solution entanglement number ((n_e)$_{soln}$). This number describes at what point the product of polymer concentration or volume fraction (ϕ_p) and molecular weight (M) of a polymer constitute fiber formation of an electrospun dope. Fiber production occurs, via intramolecular interactions, with approximately 2.5 per chain entanglements in polymer/good solvent systems.

$$(n_e)_{soln} = \frac{M_w}{(M_e)_{soln}} = \frac{\left(\phi_p M_w\right)}{M_e} \tag{1}$$

Factors of the electrospinning instrument include the needle tip to collector distance (TCD), the inner gauge diameter of the dispensing needle [3], the voltage intensity (kV) [4] and the pump feed rate. These parameters are outlined and comprehensively discussed in a review written by Haider et al. [5].

The practical focus of electrospinning is for development of filtration substrates: Such as for desalination [6] and air, water [7] or oil [8] filtration [9], nano-composite fiber reinforcements and for use as an interlayer in composite material systems like laminates and textiles. Electrospun fibers show potential in filtration because of the controlled fiber diameter, high specific strength, high surface area and mat pore sizes. However, the fiber network's mechanical instability in comparison to cast films made with the same materials prevents its use in air and liquid filtration substrates. Internal filler orientation within the submicron fibers [10], and alignment of the submicron fibers [11,12] by collection on a rotating drum, are approaches that can increase the mechanical strength of a spun mat [13]. Submicron fiber size and controlled fiber length, in combination with a nano-scale reinforcing filler, provide material rigidity and the material properties can be measured with nano-indentation [14] before compounding. Chemical crosslinking of electrospun mats by physical (heat treatment) or chemical (solvent soaking) means improves the mechanical properties within the spun mat or membrane systems. Interfacial interactions and crosslinking with sizing agents (silane) and heat treatment could be used in reinforced matrix systems or interlayer systems. Further development in these areas is necessary.

Poly (vinyl alcohol) (PVA) is a hydrophilic polymer that dissolves in water. Its desirable properties include abrasion resistance, alkali resistance, gas permeability in nonwoven fiber network, oxygen barrier properties of films, and biodegradability. As the degree of hydrolysis (DH) increases, so do its strength, water resistance and chemical compatibility [15]. Early research conducted with electrospun fiber production focused on using 87–96% hydrolyzed PVA [4,15–17]. The surface tension of the aqueous solution increased as the DH in PVA increased, such that the electric field of the electro spinning mechanism was incapable of overcoming the surface tension at 99+ DH. From 87 to 95% DH, the surface tension increased from 51 to 54 mJ/m^2 while water has a surface tension of 72 mJ/m^2. A 99% hydrolyzed PVA ($\approx$69 mJ/m^2) can be spun into fibers, but only with the help of a surfactant at concentration levels greater than or equal to 0.5% (*v/v*). The addition of Triton X-100 surfactant reduces the contact angle of 100% hydrolyzed PVA from 102.2 ± 0.4° to 60° after using 0.3 *v/w* % [16].

In desalination, a novel hydrophobic/hydrophilic (1:4 ratio) of polyethylene terephthalate (PET) and PVA interpenetrating network support was produced for a forward osmosis membrane. Improvements to flux were attributable to the wettability and water transfer capabilities of the composite support layer [18]. The use of nanofibers in air and oil filtration substrates as an added electrospun nano-layer was studied by Feng et al. [8]. Submicron fibers produced pore sizes of 14.4 µm accuracy in the nano layer, with pressure drops increasing by 3 L/min.

Customized production of electrospun fibers was studied as a bulk mechanical reinforcement in a study from Karimi et al. [19]. In this study a conventional wet lay-up method was compared to a newly developed electrospinning-electrospraying technique that utilized aligned PVA fibers as an epoxy reinforcement. It was noted that successful impregnation using vacuum consolidation and reduction of sprayed fiber handling produced results that created a shift in thermal properties, increased strength of the spun nanofiber epoxy, and increased modulus and elongation at break; results not typically seen in other nano-fillers. The increased interfacial adhesion between fiber and matrix were credited with the increased mechanical performance. Elongation improvements were proposed to be a result of percolation networks that redistributed mechanical failures in the matrix. The authors also noted that the process is scalable for automated processing.

For applications where a disturbance of spun fibers is likely, post treatments can be used to fuse fiber networks together and affect nanofiber mat properties like pore size, fiber diameter, hydrophobicity, elongation and tensile strength and stiffness. Es-saheb and Elzatahry [20] studied thermal post treatment, above and below the glass transition temperature (T_g) on PVA sheets. The authors observed a strain rate dependence of the tested samples with mechanical properties increasing with the increased strain rate. Additionally, mechanical properties were improved with heat treatments at 70% of the melting temperature (T_m). Guirguis & Moselhey [21] reported a T_g of 100% hydrolyzed 125 kg mol^{-1} PVA to be 209.6 °C. The melting temperature increases with increased DH and M_w. Miraftab et al. [22] conducted a similar heat-treatment study with the additional comparison of a methanol treatment; a similar study was done before by Franco et al. [23]. It was reported that heat treatments did not affect fiber morphology, caused yellowing or discoloration, increased water stabilization, increased crystallization, increased fiber diameter, and produced mechanically strong and stable fibers at temperatures of 180 °C for durations up to 8 h. Methanol treatment coalesced fiber morphology, leaving some pores, and improved swelling when saturated with water. In the Franco et al. study, the addition of glutaraldehyde increased elongation. Wong et al. [24] studied fiber diameters and water absorbance effect on fiber diameter with respect to four different heat treatment temperatures, ranging from 85 °C to 160 °C, at 4 h lengths, followed by 1–30 day durations of water exposure. Fiber diameters increased with heat treatment, then decreased with exposure in water. The stability of fibers, and the increase in elastic modulus, was attributed to a minimum of 4 h of treatment at 135 °C; these properties were an effect of increased crystallinity. The author also noted that increased temperatures at shorter heat exposure times would yield similar results.

Cellulose nano crystals (CNC) can be derived from numerous sources and are typically created by acid hydrolysis of native cellulose, beginning with an acid and deionized (DI) solution, followed by mechanical shearing via a centrifugation or an ultra-sonication treatment to remove amorphous regions, and ending with a washing phase [25,26]. The dimensions of CNC produced from plant sources typically range from 100–700 nm in length (L) and 5–30 nm in diameter (D) [26,27]. These sizes are suitable for utilization in submicron fiber reinforcement. The elastic modulus of CNCs were theoretically estimated to 220 ± 50 GPa [28], and measured in cotton fibers with Raman spectroscopy at 105 GPa [29]. Lee and Deng [14] employed nano-indentation to measure both the modulus of pure PVA and 20 wt% cellulose nano-whiskers reinforced nanofibers. The moduli were 2.1 and 7.6 GPa, respectively. The authors reported a linear increase with the addition cellulose nano-whiskers up to 20 wt% content. Experimental results were 60–80% higher than the isotropic Halpi-Tsai model results, but lower than the longitudinal predictions; this suggested the nano-whiskers were partially oriented within the spun fiber. Dong et al. [10] reported a 17% increase in storage modulus using nano-intentation dynamic mechanical analysis (nano-DMA) with the addition of 17 wt% CNC.

Thermal degradation of CNC begins near 200 °C, and is measured using thermogravimetric analysis (TGA) [26]. More detailed degradation data that was based on the drying method was observed and reported by Peng et al. [30]. Micro-crystalline cellulose (MCC) degraded in three steps, and was characterized by dehydration in regions 1 and 2, followed by chain depolymerization and breakdown into hydrocarbons in stage 3. Addition of CNC created a thermal shift via –OH interactions with the composite matrix. Dong et al. [10] reported a thermal shift that was attributable to hydroxyl interactions with the carbonyl groups in the poly (methyl methacrylate) matrix. Such shifts were reported in PVA/CNC electrospun nanofibers, and were attributable to hydrogen bonds between PVA side chains and CNC [13].

Cellulose and PVA composites show potential for material development due to the combined light-weight, bio-functional and environmentally friendly properties. Meree et al. reported mechanical properties with a 13% increase in strength and a 34% increase in modulus, with the addition of 3% (*w*/*w*) microcellulose. Additionally, softwood Kraft pulp has shown a doubling of tensile strength, and 2.5 times increase in the stiffness of a PVA matrix with the addition of 5% (*w*/*w*) [31].

In an effort to better understand the viscoelastic behavior of highly loaded PVA/CNC composite suspensions, rheological measurements were performed to study the composite behavior for applications requiring water-based processing. CNC loading levels of 20–67% (*w*/*w*) were combined with PVA that had an M_w of 31–50k, up to 146–186k. The increased M_w of PVA showed an increase in crystallite formation, elasticity and viscosity. As the solution aged, the rate of aging, defined by the entanglement of polymer chains in a still solution, decreased with increased time. The CNC was shown to reduce the kinetic driving force between the stratification of PVA and water in solution.

Additionally, the percolation threshold of CNC had an effect on the storage modulus (G′), in that at a lower M_w of PVA, the intramolecular entanglement played a larger elastic role as its concentration increased. However, at a high M PVA, the percolation threshold attributable to CNC is the contributing factor in rheological responses. In solution, CNC was known to show shear thinning behavior [25]. At 7% (*w*/*v*) concentration of PVA with 5, 10 and 15% (*w*/*w*) CNC the G′ was increased three times, using fully hydrolyzed PVA (99%) [17]. A lesser hydrolyzed polymer network was expected to show lower mechanical property values because of less intermolecular hydrogen bonding within the random electrospun network.

This body of work attempted to synthesize the cited literature into a repeatable method with the intent to produce consistently manufactured composite PVA/CNC nanofibers with electrospinning techniques. This study is unique, in that a highly filled PVA/CNC dope was spun and characterized by visual, thermal and mechanical means. Additionally, a post-processing treatment was utilized to investigate the heating effect on fiber morphology and mechanical properties. The target applications are for the use of these fibers as filtration media, fiber reinforcement or interlayer composite structures.

2. Materials and Methods

2.1. Sample Naming Conventions

The poly (vinyl alcohol) (PVA) concentration in distilled water was constant at 7% (*w*/*v*); therefore, it was not included in the labeling. The PVA within the dry mat was labeled as 100_{PVA} for neat mats, 80_{PVA} for 20% cellulose nanocrystals (CNC), and 50_{PVA} for 50% CNC (*w*/*w*). Similarly, the CNC content in dry mats were expressed as 20_{CNC} or 50_{CNC}. In combination, a neat PVA mat is coded 100_{PVA} and a 20% (*w*/*w*) CNC mat is coded $80_{PVA}20_{CNC}$. Heat treated samples were delineated from as-spun samples with an H label after the CNC content. For example, a neat PVA mat with heat treatment was expressed as $100_{PVA}H$, whereas an as-spun 50% (*w*/*w*) mat was expressed as $50_{PVA}50_{CNC}$.

2.2. Solution Preparation

A 99% hydrolyzed poly (vinyl alcohol) (PVA) with an M_w of 130,000 g mol^{-1} was purchased from Sigma Aldrich (St. Louis, MO, USA). PVA suspensions containing aqueous cellulose nanocrystals 12.5% (*w*/*v*) (CNC), produced by the USDA Forest Products Laboratory (FPL) (Madison, WI, USA), were dissolved in de-ionized (DI) water at approximately 90 °C for 3 hours in a silicone oil bath under magnetic stirring. DI water was chosen as not to increase processing complexity, and because the water evaporated upon fiber collection without affecting fiber morphology. Fiber dimensions were reported by the manufacturer at 5–20 nm in width and 150–200 nm in length [32]. Peng et al. [33] performed scanning electron microscope (SEM) analysis and particle size distribution of CNC; see the first figure (sections g–i) of the referenced material. The solution was kept under magnetic stirring during cooling to room temperature. After the solution was cooled, the surface tension was reduced for electrospinning by adding 0.5% (*v*/*v*) Triton X-100 M_w 80,000 (g mol^{-1}), Sigma Aldrich (St. Louis, MO, USA) surfactant. Solutions were prepared in the quantities of 7% PVA (*w*/*v*) and 0, 20 and 50% (*w*/*w*) CNC. The CNC weight was based on the dry basis of PVA. Solutions were stored in a sealed vessel containing a beaker of DI water to reduce the dehydration rate of the solutions. Upon using this storage method, solutions were kept for 2+ weeks without biological contamination. Solutions were magnetically stirred before electrospinning.

2.3. Electrospinning Parameters

Electrospun mats were produced with a NanoNC (Seoul, Korea) eS-Robot Electrospinning/spray System (Model ESR200R2P) using a single phase, 220 V, 60 Hz, with a voltage maximum of 30 kV. Th solution was dispensed through a 25 mm-long 18 G, 304 stainless steel blunt tip reusable needle, with 0.0472″ outer diameter and 0.0315″ inner diameter (IntelliSpense; Agoura Hills, CA, USA) that served as the anode. A rectangular flat plane glass collector was used for the grounding and collection substrate. Adhered to the collector, with 3M double-sided polypropylene tape, was a 22.86 × 38.1 cm sheet of 304 stainless steel, provided by NanoNC. A polytetrafluoroethylene (PTFE) sheet was cut with a 22.86 × 38.1 cm inner dimension and a 7–10 cm outer margin to direct the electric field to the collection area, effectively insulating the outside from collecting fibers. The robot feature on the machine was programmed to move at a rate of 50 mm/min along the lengthwise axis (*x*-axis), and incrementally step down 5 mm after each 70 mm pass within a 55 cm height (*y*-axis). For each mat, 1.86 mL of solution was dispensed, and the TCD was 15 cm. The TCD was determined based on precedent literature. This distance was chosen so that fiber morphology formed as opposed to a film [4,34,35]. Santos and Elchorn selected a TCD after running their own trials and selecting for best electrospun fibers [35]. The conditions, such as supplied voltage and feed rate, are changed relative to cellulose content in attempts to maximize throughput without creating droplet defects like those seen in Figure 1c,d. Mats were prepared at a 19–20 kV and 24.5–34 µL min^{-1} feed rate, as referenced in Table 1.

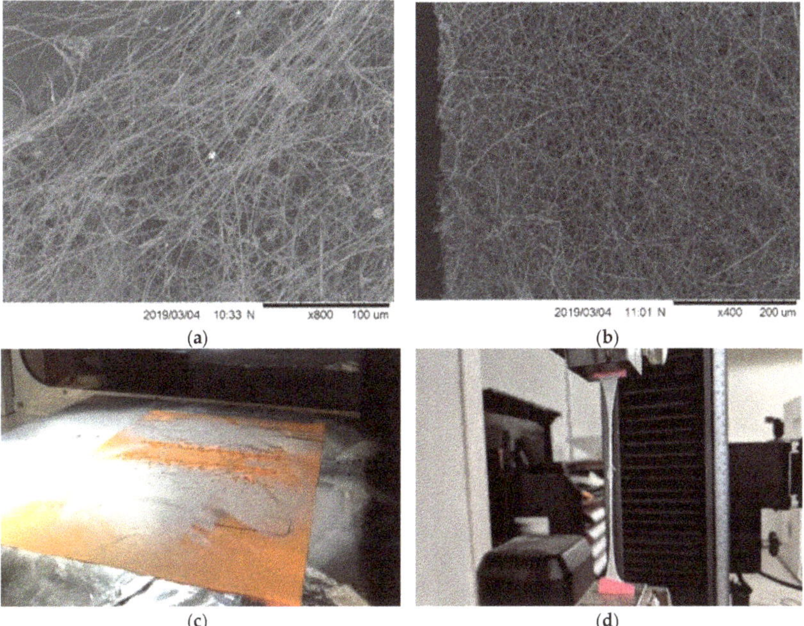

Figure 1. (**a**) $50_{PVA}50_{CNC}$ drawn specimen taken by SEM after tensile testing, (**b**) the $50_{PVA}50_{CNC}$ mat before mechanical testing, (**c**) pictured is an example of solution dispensing onto the grounded collector and producing defect in the mats, (**d**) an example of a circular film like defect produced from the dripping of solution onto the collector.

Table 1. The voltage supplied (kV) for the electrospun mats, the conductivity (µS/cm) of each solution and feed rate (µL min^{-1}) for each mat type. The sample code is explained in Section 2.1.

Sample Code	Voltage (kV)	Conductivity (µS/cm)	Feed Rate (µL min^{-1})
100_{PVA}	19–20	926.4 ± 3.1	34.0
$80_{PVA}20_{CNC}$	19–20	913.3 ± 3.3	31–32
$50_{PVA}50_{CNC}$	19–20	1037.8 ± 26.0	24.5

2.4. Scanning Electron Microscopy

Scanning Electron Microscopy (SEM) was performed using a Hitachi 2000 Tabletop Microscope SEM (Tokyo, Japan). SEM micrographs were used for fiber diameter analysis. Measurements were taken using the software ImageJ (National Institute of Health), and the measurement tool was scaled to the 10 μm scale bar of each 10,000× zoom micrograph (see images in Figure 2). Forty measurements were taken in each micrograph over three separate mats composed using the same electrospinning parameters and then averaged. One micrograph of the three mats used for measurements for each sample set is shown in Figure 1.

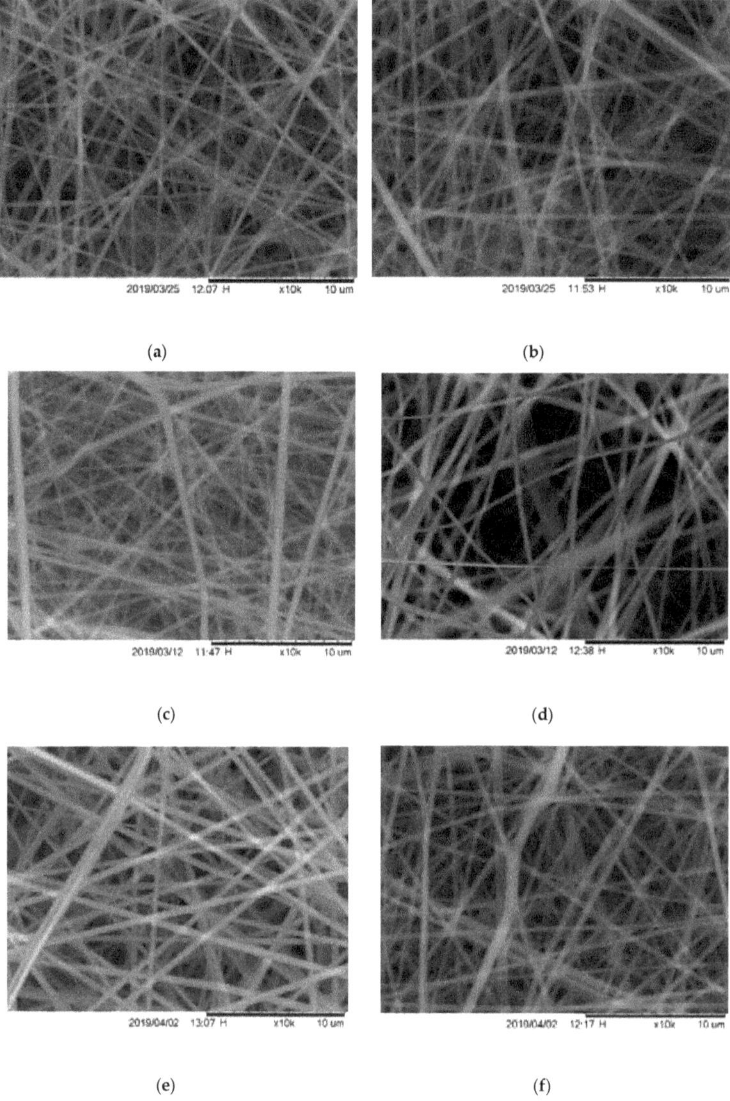

(a)

(b)

(c)

(d)

(e)

(f)

Figure 2. Scanning electron micrographs (SEM) at 10,000× magnification of random electrospun Poly (vinyl alcohol) (PVA) and PVA/cellulose nanocrystals (CNC) regular (left) and heat treated (right; H) mats. Items (**a,b**) are 100_{PVA} mats, (**c,d**) are $80_{PVA}20_{CNC}$ and (**e,f**) are $50_{PVA}50_{CNC}$.

2.5. Viscosity and Rheology

Viscosity and rheology measurements of PVA and PVA/CNC solutions were performed using a Bohlin Gemini parallel plate (25 mm) rheometer (Malvern Instruments, UK). A shear ramp at room temperature (23 °C), using shear rates ranging from 0.1 to 100 (1/s), was used to measure viscosity (Pa·s). A small amplitude oscillation test was performed with a strain sweep from 0–200% to measure the linear viscoelastic regime of each solution. A strain amplitude of 1% of the elastic modulus was chosen for frequency sweep measurements. The elastic modulus (G'), storage modulus (G'') and complex viscosity (η^*) were measured with a 0.1–100 hertz (Hz) frequency sweep. Large phase angle measurements were omitted from the reported data due to measurement lag, resulting in instrument signal noise.

2.6. Thermogravimetric Analysis (TGA)

Measurements were taken with a TA Instruments Q500 (New Castle, DE, USA). Approximately 10 mg of mat sample was used for each measurement. Samples were tested within a 30 °C to 600 °C temperature ramp with an increasing rate of 10 °C min^{-1} under nitrogen atmosphere (20 mL min^{-1}) to avoid oxidation.

Tensile testing, using a universal testing machine (Instron 5966) with a 10 kN load cell at 10 mm min^{-1} was performed using ASTM D-638 specimen IV dimensions. Three to four mats were produced for each sample category, and two samples were cut and tested from each mat. The average of the width (w), thickness (d), tensile strength and modulus were recorded. Grip surfaces were covered in adhesive paper. Samples were not conditioned, but were tested within 24 h of preparation. Heat treated samples were compressed between two PTFE sheets followed by two 0.32 cm thick aluminum sheets, and topped with a steel weight for a 2 h period at 170 °C then stored at 50% relative humidity and 23 °C.

3. Results and Discussion

3.1. SEM and Fiber Diameter

The average measured fiber diameters are reported in Table 2 and were within a range similar to cited literature with similar 99% DH and 8–15% PVA (*w/v*) solution concentrations [14,15,36]. The largest and smallest average diameters measured were $50_{PVA}50_{CNC}$ and $80_{PVA}20_{CNC}$ samples at 456 ± 85 nm and 358 ± 80 nm, respectively. The maximum and minimum measured fibers were observed in the 100_{PVA} and $80_{PVA}20_{CNC}$ sample groups at 692 nm and 233 nm, respectively.

Table 2. Average fiber diameter (nm) with standard deviation, minimum and maximum values by electrospun mat composition. The electrospun mats with heat treatment are labeled (H).

	Fiber Diameter (μm)			
Sample	**Mean**	**Stdev**	**Min**	**Max**
100_{PVA}	0.442	0.085	0.297	0.692
$100_{PVA}H$	0.390	0.077	0.254	0.586
$80_{PVA}20_{CNC}$	0.358	0.080	0.233	0.593
$80_{PVA}20_{CNC}H$	0.408	0.074	0.271	0.575
$50_{PVA}50_{CNC}$	0.456	0.078	0.272	0.632
$50_{PVA}50_{CNC}H$	0.373	0.091	0.204	0.607

Authors Lee and Deng [14] reported that fiber diameters decreased with decreased kV (10–25 kV). Fiber diameter also decreased with increasing cellulose nano-whisker content (wt.%); the TCD was reported to be 25 cm. The decrease with increasing CNC content observed in all CNC filled groups, with the exception of $50_{PVA}50_{CNC}$, coincides with the aforementioned literature. A 20 kV voltage was used in this study for numerous reasons. Primarily, a large enough voltage was needed to overcome the surface tension of the spinning dope. At lower voltages, a droplet formed between the needle tip and Taylor cone. If unaffected by the electric field the droplet fell onto the spun mat. This effect was observed for a wide range of feed rates. Unspun droplets created spot defects in the mats (Figure 1d), and potentially affected elongation properties. Secondarily, the PVA sheathe covering the CNC may have thinned as the fiber diameter decreased with heat treatment. Additionally, CNC agglomeration, or an increase in the internal filler diameter, may have further reduced the outer layer of the submicron fiber with the decrease in diameter. This phenomenon was explained in the review from Wang et al. [37].

A heat treatment regime below the T_m of PVA, but above the T_g, caused spun fibers to fuse at intersected points [20,23]. The fibers were visible in SEM micrographs, and intersections were visible as higher contrast white areas, shown in Figure 2. An especially large aggregation of randomly aligned fibers, fused together during heat treatment, is visible in the center of Figure 2d. Discoloration or yellowing was observed as a result of thermolysis [22]. The fiber diameters decreased when compared to the as-spun groups in both the 100_{PVA} and $50_{PVA}50_{CNC}$, but not the $80_{PVA}20_{CNC}$ groups; these data are reported in Table 2. The fibers did not appear more flattened or fused than the untreated mats. The preservation of fiber morphology shown in Figure 2 is consistent with Miraftab et al. [22]. Fiber diameter reduction of 10% was reported after 4 h at 135 °C in the Wong et al. [24] study. Given similar treatment conditions, the 50% (*w/w*) CNC group results reported in this study deviated slightly from the 10% reduction reported in literature, but the other groups coincided with a deviation of a few percentages. No CNC were visible at the 10,000× magnifications of SEM micrographs with a 10 µm scale bar.

3.2. Viscosity

A shear ramp at rates between 1 and 100 (s^{-1}) was applied on electrospinning dopes to obtain instantaneous viscosity (η) data (Pa·s) that are shown in Figure 3. The solution viscosity was 0.9 to 1.5 Pa·s, and exhibited similar viscosity to that reported elsewhere at high M_w (145–185 kg mol^{-1}) and with high levels of CNC filler [31]. Enayati et al. reported 209 cP at 8% (*w/v*) of PVA with a 98.9% DH and M_w of 125 kg mol^{-1} [36]. The higher viscosity response seen in the results shown in Figure 3 may be attributable to the use of parallel plates, rather than cone and plate geometries. Marginal differences in concentration and PVA M_w may have also played an additive role in the increased viscosity reported in this study.

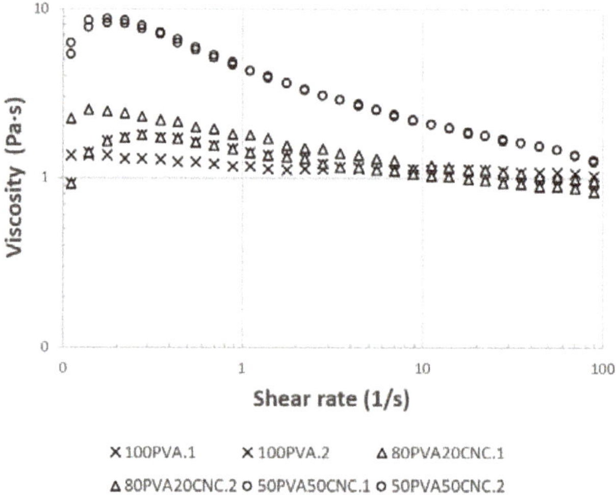

Figure 3. Instantaneous viscosity measured by a shear ramp from 1–100 (1/s). Two tests were conducted per sample type to ensure precision.

Both PVA and CNC showed shear thinning behaviors in aqueous solutions and suspension, depending on concentration. With CNC, at higher concentrations and shear rates between 0 and 1 s^{-1}, there was an initial increase in viscosity attributable to fiber network interactions. After the physical alignment of the CNC network, a more pronounced drop from 10 Pa·s was observed in 50$_{PVA}$50$_{CNC}$ samples after shear rates exceeded 0.2 (s^{-1}) [25]. In the case of PVA, hydrogen bonding of hydroxyl (–OH) groups was a major mechanism promoting shear thinning, and was affected by increased M$_w$ of polymer chains. Water and crystallites that formed by polymer chain interactions with the water solvent in standing solutions affected viscosity after 5 days [31].

Viscosity was important to measure because of its effect on fiber morphology during spinning. High viscosity may reduce pumping capacity or render the dope unable to be spun. With a shear thinning material, an increased pump pressure may cause sudden deposition of spinning dope as a result of shear forces increased inside the syringe; see Figure 1c. At high enough pressures the viscosity of the solution was decreased, and ejection of the solution occurred until the pressure was reduced. One solution to this processing issue, used in this study, was increasing the diameter of the spinning needle. Another approach was to use the relationship, presented by Shenoy et al. [2], to determine, a priori, the conditions (e.g., M$_w$) that produce fibers. Relating viscosity data for high levels of CNC to the morphology of the spun fibers is advantageous for future research of these materials.

3.3. Storage and Loss Moduli

Frequency sweep results of the storage modulus (G′), loss modulus (G″) for 100$_{PVA}$, 80$_{PVA}$20$_{CNC}$, and 50$_{PVA}$50$_{CNC}$ suspensions are shown in Figure 4. In the 100$_{PVA}$ samples, the G″ was greater than the G′, signifying the tested solution exhibited more liquid-like characteristics than solid gel-like characteristics. Additionally, the reduction of slope in the storage modulus as the CNC loading level increased suggested that the elastic dependence on frequency of G′ decreased with the increased CNC filler [31]. The attributable factor is related to the CNC percolation threshold, lack of PVA chain and crystallite formation interactions. According to the 50$_{PVA}$50$_{CNC}$ data, the suspension did not reach gelation state despite the superposition of G′ and G″ responses. These data coincided with the feed rate data reported in Table 1. Lower feed rates were necessary as the electrospinning dope reached more gelatinous states. However, spinning was possible with the aid of the Triton X-100 surfactant.

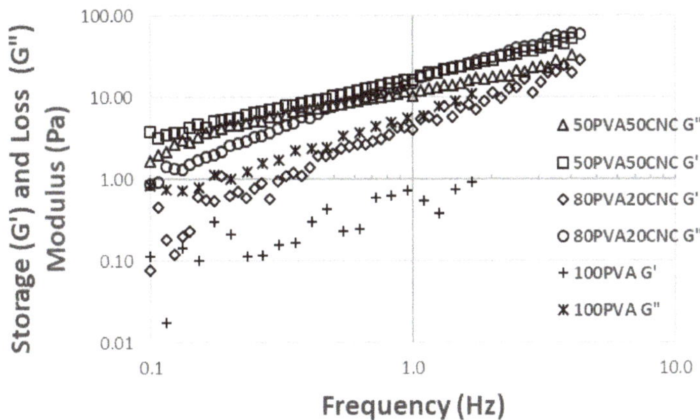

Figure 4. Data measured from a frequency (Hz) sweep (0.1–10) of aqueous suspension neat PVA and PVA/CNC at varying weight percentages. Some data was omitted because the phase angle was too large.

3.4. Complex Viscosity

Complex viscosity (η^*) was measured in a frequency sweep along with the G′ and G″ and is shown in Figure 5. The response of CNC is clearly visible between the two $80_{PVA}20_{CNC}$ and $50_{PVA}50_{CNC}$ and the 100_{PVA} sample groups. The neat samples showed an independence in response to increased frequency (Hz). The increase in η^* in was attributable to the CNC domains acting in conjunction with the liquid crystalline interphase of the PVA dope. These domains resisted forces during the frequency sweep and resulted in some dependence as frequency increased [31]. Conversely, once the PVA in the dope no longer acted as the primary resisting component, the $50_{PVA}50_{CNC}$ sample group showed an increase in η^* compared to $80_{PVA}20_{CNC}$ and 100_{PVA}. A decline was observed afterwards with the increased frequency. As the CNC networks were broken or weakened, the resistance decreased as a response to the increase in frequency, and met with $80_{PVA}20_{CNC}$ samples near 1 Hz.

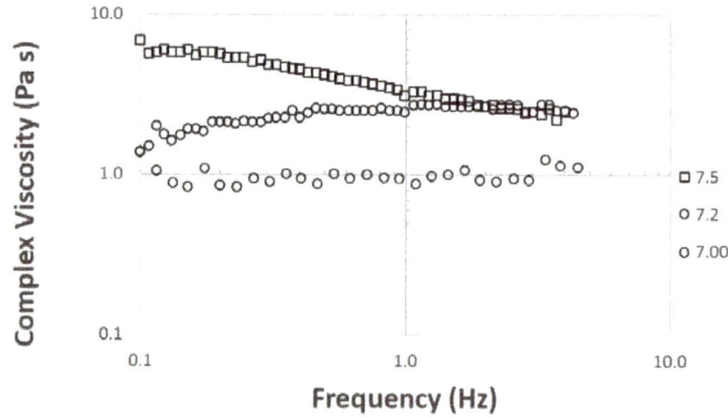

Figure 5. Complex viscosity data measured in a frequency sweep (0.1–10 Hz).

3.5. Thermogravimetric Analysis

Neat PVA and PVA/CNC mats were degraded via TGA, and the weight (%) and derivative weight loss (°C/%) data are shown in Figures 6 and 7, respectively. The neat data showed a distinct mass loss that was attributable to water starting below 100 °C in both figures. The decomposition of PVA followed a two-stage process. The onset of PVA degradation occurred at 175 °C, and continued in the range of 240–350 °C, which agrees with non-thermally treated, 99% DH results from the reported literature. The first significant stage occurred at 240–340 °C and the second at 390–470 °C [38]. The maximum rate loss (%/°C) occurred at ≈216 °C, dipped slightly and was sustained until 300 °C. This loss was attributable to PVA chain stripping and chain scission. The final loss was attributable to chain decomposition and volatilization of residual elements of the polymer backbone [17,20].

Cellulose degradation occurred in three stages. Freeze dried and spray dried CNC sourced from the FPL was characterized and reported by Peng et al. [30]. Results reported in the Peng et al. study should exhibit an accurate representation for the TG and dTG information regarding CNC material used in this study. The first region ranged from 25 °C to 216 °C, and the majority of mass loss was attributed to water evaporation. The second region ranged from the onset of thermal degradation at the end of the first to the dehydration of cellulose, occurring between 200–280 °C, followed by the de-polymerization of cellulose, occurring between 280–340 °C, with some simultaneous overlap with dehydration. Region three occurred at 358 °C and after 500 °C, where the levoglucosan decomposed into hydrocarbons and hydrogen, leaving char residue at the final temperature of 600 °C.

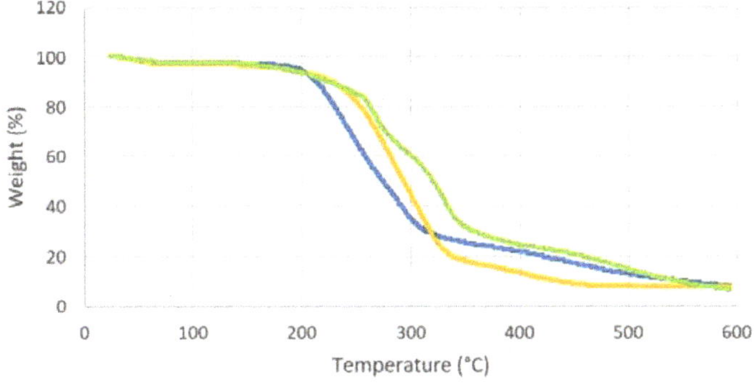

Figure 6. Thermogravimetric analysis (TGA) curves showing weight loss (%) of 100_{PVA}, $80_{PVA}20_{CNC}$ and $50_{PVA}50_{CNC}$ electrospun random mats.

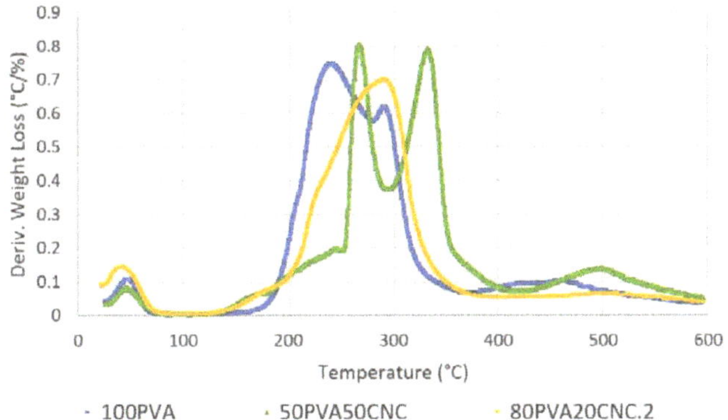

Figure 7. TGA curves showing derivative weight loss (%/°C) 100_{PVA}, $80_{PVA}20_{CNC}$ and $50_{PVA}50_{CNC}$ electrospun random composite mats.

Samples containing PVA and CNC decomposed similarly to neat PVA, but the presence of CNC shifted the peak mass loss rate (%/°C) to higher temperatures. In electrospun poly (vinylidene fluoride) fibers, studied by Zhang et al., a similar shift was observed in the 250–330 °C cellulose decomposition region in the TGA data [39]. The gradual rate increase after 125 °C with a sharp increase at ≈200–250 °C were attributable to the free water loss of CNC and the onset chain stripping of PVA. Two distinct peaks of mass loss were then visible afterward in $50_{PVA}50_{CNC}$, but appeared as one peak with a slight shoulder in the $80_{PVA}20_{CNC}$ samples. This combination effect, attributable to hydrogen bonding interactions, showed a more combined effect with the PVA at the 20% (*w/w*) CNC loading level. However, for $50_{PVA}50_{CNC}$ the first peak, shown in Figure 6, began at ≈216 °C, and dipped significantly at ≈290 °C. In this region, dehydration of CNC bound water and the de-polymerization of CNC occurred during thermal decomposition [25,30]. A second peak was attributable to CNC dehydration/de-polymerization, coupled with the chain scission of PVA, and occurred near 340 °C. The rate of decomposition slowly declined until 400 °C. The rate loss increased afterward and peaked near 500 °C. The final rate increase was attributable to the breakdown and volatilization of CNC and PVA backbone structures.

3.6. Mechanical Properties

Dimensional analysis, tensile strength and modulus of elasticity data are shown in Table 3, and were reported with mean and coefficient of variance (COV). The COV was above 20% for all of the strength and modulus data. Dimensions were under 20% COV in all samples except the $50_{PVA}50_{CNC}$ thickness results. The low COV in dimensional analysis indicated that the electrospun production method, with high percentages of CNC (w/w) filler, was repeatable. The assumption in mechanical property variance was mostly attributable to the materials performance with internal defects creating variance rather than the dimensional instability of the tested specimens. Stress strain curves are shown by sample group in Figure 8.

Table 3. Tensile mechanical properties for electrospun PVA (7% w/v) and PVA/CNC composite mats with and without heat treatments. The mean and coefficient of variation are reported for all three types (00, 20, 50% w/w) and for no treatment and heat treatment (H).

	Thickness (mm)		Width (mm)		Modulus (MPa)		Tensile Strength (MPa)	
Sample	Mean	COV	Mean	COV	Mean	COV	Mean	COV
100_{PVA}	0.09	9%	6.11	2%	371.6	35%	6.27	40%
$100_{PVA}H$	0.09	16%	6.08	2%	1645.1	29%	4.26	22%
$80_{PVA}20_{CNC}$	0.09	15%	6.41	2%	300.7	51%	5.28	30%
$80_{PVA}20_{CNC}H$	0.09	25%	6.10	4%	690.7	70%	5.08	28%
$50_{PVA}50_{CNC}$	0.08	22%	6.12	2%	785.4	56%	1.77	32%
$50_{PVA}50_{CNC}H$	0.085	10%	6.25	2%	1570.4	37%	2.04	59%

Tensile strength results in this study were similar to a study, using 0.1% PVPA and 12% (w/v) PVA with a 2, 10 and 24 h 150 °C post-heat treatment, performed by Franco et al. [23]. The tensile strength among groups declined with the addition of CNC and after heat treatment in all sample groups. Stress-strain curves are shown in Figure 8. This reduction in strength could be attributable to the overfilling of the interior of PVA fibers. Particle size played a large role in composite performance and affected stress concentration in the matrix. With increased filler content, a decrease in tensile strength is observed (see Figure 8a,c,e). The effect may be especially prevalent when spun fiber diameters are 300–400 nm in width, and the filler diameter was over half of that dimension [37]. In this particular case, if the CNC filler that was 5–30 nm in width, and then agglomerated within the core of the PVA matrix fiber, the combined widths of the CNC quickly became a significant portion of the electrospun fiber's internal width. The agglomerated CNC diameter would then reduce the tensile strength by reducing the active surface area available for interfacial interactions like hydrogen bonding. The largest averaged tensile strength was observed in 100_{PVA} at 6.27 MPa and the lowest was 1.77 MPa in $50_{PVA}50_{CNC}$ mats. After heat treatment the average tensile strength of all sample groups decreased. The largest decline, a 32% difference, was observed in $100_{PVA}H$ mats. This loss was attributable to the fiber intersection fusion. This fusion of intersecting fibers significantly reduced the strain (%), see Figure 8a and b. With the mechanical testing rate dependence of spun mats and decreased strain at yield, lower tensile strength values at the same cross head test speed were observed. Similar results were reported by Es-saheb et al. [20]. The tensile strength increased between $50_{PVA}50_{CNC}$ and $50_{PVA}50_{CNC}H$ mats by 15%. This is attributable to the diffusion of fibers intersections and the inability of the fibers to be drawn. This, however, does not imply that the fiber was an effective strength reinforcing agent in the PVA matrix, since the 50_{CNC} values were significantly lower than the neat PVA.

Modulus of elasticity data are shown in Table 3, and the mean and COV are reported. Standard deviation was over 20% in all sample groups, but were lowest in $100_{PVAC}H$ and $50_{PVA}50_{CNC}$ sample groups; these data are not shown because of the high COV displayed in each sample group. The correlation of a modulus of elasticity increase with increased CNC content (w/w) was less distinguishable between 100_{PVA} at 371.6 MPa, when compared to $80_{PVA}20_{CNC}$ samples at 300.7 MPa. However, the

785.4 MPa modulus of elasticity in $50_{PVA}50_{CNC}$ samples increased 211% with the addition of CNC. Defects in 100_{PVA} samples were attributable to droplets falling onto the spun mat. The unspun dope diffused the fibers, and may have promoted an artificial stiffening of samples, as evidenced by the shortened elongation of two samples; see the stress-strain curves shown in Figure 8a. At a 20% fiber loading level, CNC showed hydrogen bonding, as evidenced by the single dTG degradation peak seen in Figure 7. However, tensile stiffness was not increased as a result. The lack of hydrogen bonding may be attributable to decreased crystallization of PVA with the addition of additives [36]. The crystalline domain of the PVA interacting with the CNC may have created discontinuity within the spun fibers and decreased stiffness [10]. A stronger response was seen in the $50_{PVA}50_{CNC}$ samples, and stiffness increased as CNC may have created a percolation effect within the spun fiber.

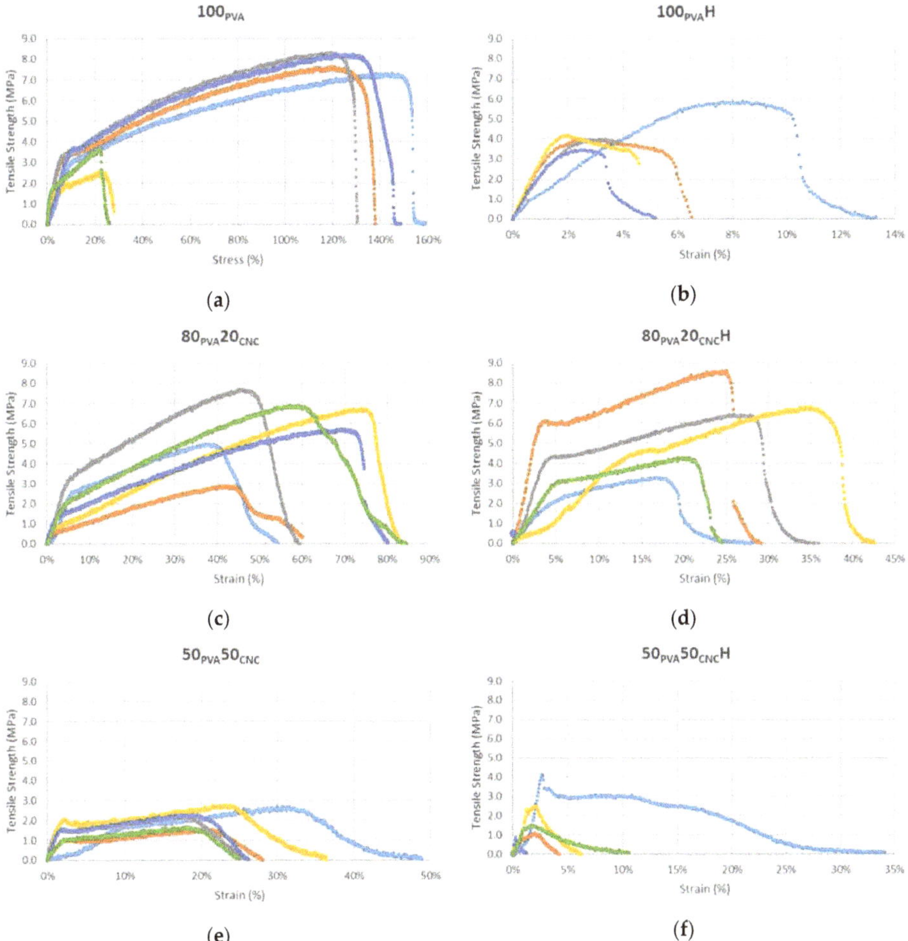

Figure 8. Stress-strain curves for 100_{PVA} (**a,b**), $80_{PVA}20_{CNC}$ (**c,d**) and $50_{PVA}50_{CNC}$ (**e,f**) tensile strength samples. Each sample group has six samples (with the exception of 100_{PVA}), and heat treated samples are labeled with an (H). Colors are used to differentiate between individual samples *within* each group category.

With regard to heat treatment, in 100_{PVA}H samples the tensile stiffness was highest at 1656.1 MPa in 100_{PVA}. Fusion between fibers and molecular entanglement are factors contributing to the increase in stiffness. Strain (%) reduction and increased modulus of the mat appeared to be correlated, as seen in stress-strain graphs shown in Figure 8. The second strongest modulus of elasticity was 1570.4 MPa in $50_{PVA}50_{CNC}$H, a 5% decrease from 100_{PVA}. The crystallization and diffusion of PVA fibers coupled with half CNC (*w/w*) filler increased the $50_{PVA}50_{CNC}$H stiffness. The 5% reduction, compared to 100_{PVA}, may be attributable to a discontinuity of the matrix attributable to clogging during processing or the lack of PVA CNC adhesion described earlier.

Stress-strain data is reported in Figure 8. The higher strain percentages were attributable to a lengthening of the random mat and an aligning of the fibers during drawing; see Figure 1a,b. Fusion of the electrospun layers occurred as a result of the liquid droplet defects. The fused areas created stress concentration points and prohibited drawing of fibers under tensile load. A circular shape is visible in the middle of the sample shown in the specimen pictured in Figure 1d. This phenomenon effectively shortened the extension potential of the mat, and resulted in variation among other 100_{PVA} samples. This shortening effect is visible in the stress strain curves of 100_{PVA} and $80_{PVA}20_{CNC}$ in Figure 8a,c, respectively. Reduction in drop defects may reduce the strain defects seen in decreased strain (%) samples.

4. Conclusions

This work concludes that electrospinning of highly filled (up to 50% *w/w*) PVA/CNC composite material was possible with the chosen electrospinning parameters. Consistent morphology between as-spun mats was observed in comparison to heat treated sample groups by SEM. The fiber diameters of electrospun random mats were decreased in 100_{PVA} and $50_{PVA}50_{CNC}$ by 12 and 18% as a result of PVA crystallization during heat treatment. The presence of CNC fibers in random PVA electrospun mats were confirmed by an increased thermal stability in $80_{PVA}20_{CNC}$ and $50_{PVA}50_{CNC}$ sample groups, increasing the materials' thermal use range. The $80_{PVA}20_{CNC}$ sample group decomposed in a homogenous fashion, while the $50_{PVA}50_{CNC}$ decomposed into a two peak, three phase, pattern. The former degradation pattern was attributable to improved hydrogen bonding. The modulus of elasticity of 100_{PVA} mats increased by 211% as a result of 50% (*w/w*) added CNC filler, and may be attributable to percolation networks not observed in 20% (*w/w*) CNC stiffness results. This increase in tensile stiffness of as-spun submicron fibers could be applied as a means to fiber reinforce a polymeric resin matrix. Additionally, in comparison to as-spun samples, heat treatment of mats for 2 h at 170 °C increased tensile stiffness by 4.4-, 2.3- and 2-fold in 100_{PVA}H, $80_{PVA}20_{CNC}$H and $50_{PVA}50_{CNC}$H, respectively. Decreased tensile strength and shortened strain at yield were observed with the addition of increased CNC content and heat treatment. In PVA and PVA/CNC sample groups the heat treatment promoted physical diffusion in the fiber mat and the increase in stiffness without noticeable morphological changes. These attributes could be utilized in combination with hydrophobic filtration media, and employed as an interlayer reinforcement of composite laminate materials.

Author Contributions: Conceptualization, D.J.G., Y.H. and J.E.S.; methodology, J.E.S., D.J.G. and Y.H.; software, J.E.S.; validation, D.J.G., Y.H. and T.S.R.; formal analysis, J.E.S.; investigation, J.E.S.; resources, D.J.G., T.S.R.; data curation, J.E.S.; writing—original draft preparation, J.E.S.; writing—review and editing, D.J.G., T.S.R.; visualization, J.E.S.; supervision, D.J.G., T.S.R. and Y.H.; project administration, D.J.G.; funding acquisition, D.J.G.

Funding: The work of the authors was sponsored by the U.S. Army Engineer Research and Development Center (ERDC) under CEED-17-0018 "Engineered Energy Efficient and Low Logistic Burden Materials and Processes" executed under Contract Number W15QKN-17-9-8888. Permission to publish was granted by the director of the ERDC Geotechnical and Structures Laboratory.

Acknowledgments: The testing was conducted, with support from the students and staff, at the Advanced Structures and Composites Center and the School of Forest Resources at the University of Maine, Orono, ME 04469, USA.

Conflicts of Interest: The authors declare no conflict of interest.

References

1. Laudenslager, M.J.; Sigmund, W.M. Electrospinning. In *Encyclopedia of Nanotechnology*; Bhushan, B., Ed.; Springer: Dordrecht, The Netherlands, 2012; pp. 769–775. ISBN 978-90-481-9751-4.
2. Shenoy, S.L.; Bates, W.D.; Frisch, H.L.; Wnek, G.E. Role of chain entanglements on fiber formation during electrospinning of polymer solutions: Good solvent, non-specific polymer-polymer interaction limit. *Polymer* **2005**, *46*, 3372–3384. [CrossRef]
3. Macossay, J.; Marruffo, A.; Rincon, R.; Eubanks, T.; Kuang, A. Effect of needle diameter on nanofiber diameter and thermal properties of electrospun poly(methyl methacrylate). *Polym. Adv. Technol.* **2007**, *18*, 180–183. [CrossRef]
4. Ding, B.; Kim, H.Y.; Lee, S.C.; Shao, C.L.; Lee, D.R.; Park, S.J.; Kwag, G.B.; Choi, K.J. Preparation and characterization of a nanoscale poly(vinyl alcohol) fiber aggregate produced by an electrospinning method. *J. Polym. Sci. Part. B Polym. Phys.* **2002**, *40*, 1261–1268. [CrossRef]
5. Haider, A.; Haider, S.; Kang, I.K. A comprehensive review summarizing the effect of electrospinning parameters and potential applications of nanofibers in biomedical and biotechnology. *Arab. J. Chem.* **2015**, *11*, 1165–1188. [CrossRef]
6. Woo, Y.C.; Tijing, L.D.; Shim, W.G.; Choi, J.S.; Kim, S.H.; He, T.; Drioli, E.; Shon, H.K. Water desalination using graphene-enhanced electrospun nanofiber membrane via air gap membrane distillation. *J. Membr. Sci.* **2016**, *520*, 99–110. [CrossRef]
7. Balamurugan, R.; Sundarrajan, S.; Ramakrishna, S. Recent trends in nanofibrous membranes and their suitability for air and water filtrations. *Membranes* **2011**, *1*, 232–248. [CrossRef] [PubMed]
8. Feng, J.Y.; Zhang, J.C.; Yang, D. Preparation and Oil Filtration Properties of Electrospun Nanofiber Composite Material. *J. Eng. Fibers Fabr.* **2014**, *9*. [CrossRef]
9. Kenry; Lim, C.T. Nanofiber technology: current status and emerging developments. *Prog. Polym. Sci.* **2017**, *70*, 1–17. [CrossRef]
10. Dong, H.; Strawhecker, K.E.; Snyder, J.F.; Orlicki, J.A.; Reiner, R.S.; Rudie, A.W. Cellulose nanocrystals as a reinforcing material for electrospun poly(methyl methacrylate) fibers: Formation, properties and nanomechanical characterization. *Carbohydr. Polym.* **2012**, *87*, 2488–2495. [CrossRef]
11. Pan, H.; Li, L.; Hu, L.; Cui, X. Continuous aligned polymer fibers produced by a modified electrospinning method. *Polymer* **2006**, *47*, 4901–4904. [CrossRef]
12. Herrera, N.V.; Mathew, A.P.; Wang, L.Y.; Oksman, K. Randomly oriented and aligned cellulose fibres reinforced with cellulose nanowhiskers, prepared by electrospinning. *Plast. Rubber Compos.* **2011**, *40*, 57–64. [CrossRef]
13. López de Dicastillo, C.; Garrido, L.; Alvarado, N.; Romero, J.; Palma, J.; Galotto, M. Improvement of Polylactide Properties through Cellulose Nanocrystals Embedded in Poly(Vinyl Alcohol) Electrospun Nanofibers. *Nanomaterials* **2017**, *7*, 106. [CrossRef]
14. Lee, J.; Deng, Y. Nanoindentation study of individual cellulose nanowhisker-reinforced PVA electrospun fiber. *Polym. Bull.* **2013**, *70*, 1205–1219. [CrossRef]
15. Park, J.-C.; Ito, T.; Kim, K.-O.; Kim, K.-W.; Kim, B.-S.; Khil, M.-S.; Kim, H.-Y.; Kim, I.-S. Electrospun poly(vinyl alcohol) nanofibers: effects of degree of hydrolysis and enhanced water stability. *Polym. J.* **2010**, *42*, 273–276. [CrossRef]
16. Yao, L.; Haas, T.W.; Guiseppi-Elie, A.; Bowlin, G.L.; Simpson, D.G.; Wnek, G.E. Electrospinning and Stabilization of Fully Hydrolyzed Poly(Vinyl Alcohol) Fibers. *Chem. Mater.* **2003**, *15*, 1860–1864. [CrossRef]
17. Habibi, Y.; Zoppe, J.O.; Rojas, O.J.; Peresin, M.S.; Pawlak, J.J. Nanofiber Composites of Polyvinyl Alcohol and Cellulose Nanocrystals: Manufacture and Characterization. *Biomacromolecules* **2010**, *11*, 674–681.
18. Tian, E.L.; Zhou, H.; Ren, Y.W.; Wang, X.Z.; Xiong, S.W. Novel design of hydrophobic / hydrophilic interpenetrating network composite nano fi bers for the support layer of forward osmosis membrane. *DES* **2014**, *347*, 207–214. [CrossRef]
19. Karimi, S.; Staiger, M.P.; Buunk, N.; Fessard, A.; Tucker, N. Uniaxially aligned electrospun fibers for advanced nanocomposites based on a model PVOH-epoxy system. *Compos. Part A Appl. Sci. Manuf.* **2016**, *81*, 214–221. [CrossRef]
20. Es-saheb, M.; Elzatahry, A. Post-Heat Treatment and Mechanical Assessment of Polyvinyl Alcohol Nanofiber Sheet Fabricated by Electrospinning Technique. *Int. J. Polym. Sci.* **2014**, *2014*, 1–6. [CrossRef]

21. Guirguis, O.W.; Moselhey, M.T.H. Thermal and structural studies of poly (vinyl alcohol) and hydroxypropyl cellulose blends. *Nat. Sci.* **2012**, *4*, 57–67. [CrossRef]

22. Miraftab, M.; Saifullah, A.N.; Çay, A. Physical stabilisation of electrospun poly(vinyl alcohol) nanofibres: comparative study on methanol and heat-based crosslinking. *J. Mater. Sci.* **2015**, *50*, 1943–1957. [CrossRef]

23. Franco, R.A.; Min, Y.K.; Yang, H.M.; Lee, B.T. On stabilization of PVPA/PVA electrospun nanofiber membrane and its effect on material properties and biocompatibility. *J. Nanomater.* **2012**, *2012*. [CrossRef]

24. Wong, K.K.H.; Zinke-Allmang, M.; Wan, W. Effect of annealing on aqueous stability and elastic modulus of electrospun poly(vinyl alcohol) fibers. *J. Mater. Sci.* **2010**, *45*, 2456–2465. [CrossRef]

25. Moon, R.J.; Martini, A.; Nairn, J.; Simonsen, J.; Youngblood, J. Cellulose nanomaterials review: structure, properties and nanocomposites. *Chem. Soc. Rev.* **2011**, *40*, 3941–3994. [CrossRef]

26. Lin, N.; Huang, J.; Dufresne, A. Preparation, properties and applications of polysaccharide nanocrystals in advanced functional nanomaterials: A review. *Nanoscale* **2012**, *4*, 3274–3294. [CrossRef]

27. Beck-Candanedo, S.; Roman, M.; Gray, D.G. Effect of reaction conditions on the properties and behavior of wood cellulose nanocrystal suspensions. *Biomacromolecules* **2005**, *6*, 1048–1054. [CrossRef] [PubMed]

28. Diddens, I.; Murphy, B.; Krisch, M.; Müller, M. Anisotropic elastic properties of cellulose measured using inelastic X-ray scattering. *Macromolecules* **2008**, *41*, 9755–9759. [CrossRef]

29. Rusli, R.; Eichhorn, S.J. Determination of the stiffness of cellulose nanowhiskers and the fiber-matrix interface in a nanocomposite using Raman spectroscopy. *Appl. Phys. Lett.* **2008**, *93*, 033111. [CrossRef]

30. Peng, Y.; Gardner, D.J.; Han, Y.; Kiziltas, A.; Cai, Z.; Tshabalala, M.A. Influence of drying method on the material properties of nanocellulose I: Thermostability and crystallinity. *Cellulose* **2013**, *20*, 2379–2392. [CrossRef]

31. Meree, C.E.; Schueneman, G.T.; Meredith, J.C.; Shofner, M.L. Rheological behavior of highly loaded cellulose nanocrystal/poly(vinyl alcohol) composite suspensions. *Cellulose* **2016**, *23*, 3001–3012. [CrossRef]

32. Product Specification. Available online: https://umaine.edu/pdc/wp-content/uploads/sites/398/2016/03/Specs-CNF.pdf (accessed on 24 May 2019).

33. Peng, Y.; Gardner, D.J.; Han, Y. Drying cellulose nanofibrils: in search of a suitable method. *Cellulose* **2012**, *19*, 91–102. [CrossRef]

34. Habibi, Y.; Lucia, L.A.; Rojas, O.J. Cellulose Nanocrystals: Chemistry, Self-Assembly, and Applications. *Chem. Rev.* **2010**, *110*, 3479–3500. [CrossRef]

35. Wanasekara, N.D.; Santos, R.P.O.; Douch, C.; Frollini, E.; Eichhorn, S.J. Orientation of cellulose nanocrystals in electrospun polymer fibres. *J. Mater. Sci.* **2016**, *51*, 218–227.

36. Enayati, M.S.; Behzad, T.; Sajkiewicz, P.; Bagheri, R.; Ghasemi-Mobarakeh, L.; Kuśnieruk, S.; Rogowska-Tylman, J.; Pahlevanneshan, Z.; Choińska, E.; Święszkowski, W. Fabrication and characterization of electrospun bionanocomposites of poly (vinyl alcohol)/nanohydroxyapatite/cellulose nanofibers. *Int. J. Polym. Mater. Polym. Biomater.* **2016**, *65*, 660–674. [CrossRef]

37. Wang, G.; Yu, D.; Kelkar, A.D.; Zhang, L. Electrospun nanofiber: Emerging reinforcing filler in polymer matrix composite materials. *Prog. Polym. Sci.* **2017**, *75*, 73–107. [CrossRef]

38. Yang, H.; Xu, S.; Jiang, L.; Dan, Y. Thermal decomposition behavior of poly (vinyl alcohol) with different hydroxyl content. *J. Macromol. Sci. Part B Phys.* **2012**, *51*, 464–480. [CrossRef]

39. Zhang, Z.; Wu, Q.; Song, K.; Lei, T.; Wu, Y. Poly(vinylidene fluoride)/cellulose nanocrystals composites: rheological, hydrophilicity, thermal and mechanical properties. *Cellulose* **2015**, *22*, 2431–2441. [CrossRef]

 *nanomaterials*

Article

Improved Dispersion of Bacterial Cellulose Fibers for the Reinforcement of Paper Made from Recycled Fibers

Zhouyang Xiang [1], Jie Zhang [1], Qingguo Liu [2], Yong Chen [2], Jun Li [1] and Fachuang Lu [1,3,*]

[1] State Key Laboratory of Pulp and Paper Engineering, South China University of Technology, Guangzhou 510640, China; fezyxiang@scut.edu.cn (Z.X.); zhangjie@brunp.com.cn (J.Z.); ppjunli@scut.edu.cn (J.L.)

[2] Nanjing High Tech University Biological Technology Research Institute Co., Ltd., Nanjing 211899, China; liuqingguo@njiwb.com (Q.L.); chenyong@njiwb.com (Y.C.)

[3] Guangdong Engineering Research Center for Green Fine Chemicals, Guangzhou 510640, China

* Correspondence: fefclv@scut.edu.cn; Tel.: +86-20-87113953

Received: 28 November 2018; Accepted: 29 December 2018; Published: 4 January 2019

Abstract: Bacterial cellulose (BC) can be used to improve the physical properties of paper. However, previous studies have showed that the effectiveness of this improvement is impaired by the agglomeration of the disintegrated BC fibers. Effective dispersion of BC fibers is important to their reinforcing effects to paper products, especially those made of recycled fibers. In this study, carboxymethyl cellulose, xylan, glucomannan, cationized starch, and polyethylene oxide were used to improve the dispersion of BC fibers. With dispersed BC fibers, the paper made of recycled fiber showed improved dry tensile strength. The best improvement in dry tensile index was 4.2 N·m/g or 12.7% up, which was obtained by adding BC fibers dispersed with glucomannan. Glucomannan had the highest adsorption onto BC fibers, i.e., 750 mg/g at 1000 mg/L concentration, leading to the best colloidal stability of BC fiber suspension that had no aggregation in 50 min at 0.1 weight ratio of glucomannan to BC. TEMPO-mediated oxidation of BC was effective in improving its colloidal stability, but not effective in improving the ability of BC fiber to enhance paper dry tensile index while the wet tensile index was improved from 0.89 N·m/g to 1.59 N·m/g, i.e., ~80% improvement.

Keywords: bacterial cellulose; dispersion; recycled fiber; reinforcement; tensile strength

1. Introduction

Due to the low-carbon environmental protection concept, a large proportion of paper and paperboard produced are recycled every year. In 2014, approximately 60% of the raw materials for pulp and paper industries were fibers recycled from waste paper and paperboard globally [1]. However, paper or paperboard made of recycled fibers generally had inferior physical properties compared to those made of virgin fibers [2–4]. Consequently, effective reinforcing agents are used for processing recycled fibers into quality paper products. In addition, paper with excessive inorganic fillers for special surface properties, e.g., fire retardant, also requires reinforcement [5]. Nano-cellulose, due to their structural similarity, compatibility and affinity to pulp fibers, as well as their high mechanical strength, was used to reinforce paper made from recycled fibers [3,6–8] or paper with excessive inorganic fillers [5].

Bacterial cellulose (BC) is a special type of nano-cellulose secreted in vitro by bacteria, of which the most studied species is *Gluconacetobacter xylinus* [9]. BC has the same chemical structure as plant-based cellulose and has a higher crystallinity and degree of polymerization [10,11]. The microstructure of BC is also different from plant-based cellulose. BC does not have a macrofibril structure. The BC microfibrils are 10–100 nm in diameter and interlace with each other forming a fine net structure [12–15], leading to a high specific surface area. When BC pellicles are disintegrated into small fragments or fibers, abundant free hydroxyls will be released. Being used as reinforcing agents during paper-making, the BC fibers can bridge between fibers and improve the physical properties of papers. Our previous studies have already shown that the addition of BC lower than 1% of paper dry weight can effectively improve the tensile strength of paper produced from some high-quality fibers, such as softwood pulp, hardwood pulp, sugarcane bagasse pulp, and bamboo pulp; these fibers are more fibrillated and, thus, have more surface hydrogen bonding sites for BC fibers bridging between [12]. However, the reinforcing effect of BC on recycled fiber paper was limited [12]. Recycled fibers suffer from fiber hornification causing less surface hydrogen bonding sites [3,6,7,12]. Nevertheless, BC fibers with abundant free hydroxyls should have been able to increase the total numbers of hydrogen bonding sites within the recycled fiber matrix and thus the paper strength [3,6,7]. However, BC fibers aggregate with themselves instead of being evenly distributed within the fiber matrix, causing little improvements in the tensile strength of paper produced from recycled fibers [12]. The agglomeration was also a problem when using nano-cellulose, e.g., cellulose nanofibers (CNF) and nanocrystals (CNC), as reinforcing agents for recycled fiber paper [3].

In addition to the mixing of nano-cellulose with plant fibers during sheet forming, the coating of inorganic nanoparticles onto paper surface may also be an effective method to reinforce paper. The coating of $Mg(OH)_2$ nanoparticles [16] and Halloysite ($Al_2Si_2O_5(OH)_4 \cdot 2H_2O$) nanotubes [17,18] onto paper surface can improve paper tensile strength by 40–50%. The dispersion and stabilization of those inorganic nanoparticles were vital to their reinforcing effects; e.g., the $Mg(OH)_2$ nanoparticles were stabilized trimethylsilyl cellulose [16] and the halloysite nanotubes were stabilized by hydroxypropylcellulose [17,18].

Consequently, the dispersion or stabilization of nano-materials was one of the key factors in determining their effects in paper reinforcement. This study is to explore the effective dispersion or stabilization of BC fibers so that their reinforcing effects on paper made from recycled fibers could be improved. Two types of stabilization mechanisms, i.e., steric and electrostatic repulsions, are often used to explain the dispersion or the colloidal stability of cellulose nanofibers or nanocrystals suspended in water [2]. A good dispersion or colloidal stability can be reached by non-covalently adsorbing or covalently grafting macromolecules to the fiber surface, which prevents the fibers from aggregating due to steric hindrance [19]. Many natural polymers or macromolecules are water soluble and have good adhesion to cellulose fibers such as xyloglucan, β-D-glucan, carboxymethyl cellulose (CMC), etc. [19,20]. Having good adhesion to cellulose fibers, xyloglucan had been proved to disperse cellulose nanocrystals well [19,21]. Adding CMC to cellulose fiber suspension led to better dispersion than adding xyloglucan, which may be due to the surface charges of CMC [19]. Bulky macromolecules or polymer chains, for instance, poly(ethylene oxide) or poly(phenylene oxide), can also be grafted onto cellulose fiber (nanocrystalline cellulose) surfaces leading to the formation of well-dispersed and stable colloidal aqueous suspension [22,23]. The hydroxyls on cellulose nanofiber surface may be chemically modified forming charged groups on surface so that the electrostatic repulsion effects prevent such fibers from aggregating [24]. Cellulose nanofibers or nanocrystals produced by TEMPO-mediated oxidation or concentrated H_2SO_4 treatment would have their surfaces modified with negative charges, which may naturally be good for their dispersion [2]. However, for general cellulose nanofibers, surface modification with anionic agents is usually required to obtain better dispersions. Meanwhile, cationic surface modification can also improve the dispersion of cellulose fibers [25]. Cellulose nanofibers can be grafted with glycidyltrimethylammonium chloride (GTMAC) [25] or epoxypropyltrimethylammonium chloride [26] to improve their dispersions.

In addition to effective dispersion, the increased retention of nano-cellulose during paper forming may further improve the paper strength. With addition of cationic macromolecules, e.g., chitosan and cationic polyacrylamide, CNC and CNF can more effectively improve the strength of paper produced from recycled fibers [3]. The cationic modification of BC fibers increased their retention in paper and, thus, further improved the strength of paper made of sugarcane bagasse pulp [13]. However, BC fiber retention is not a problem for recycled fiber. Our previous study has shown that the BC fiber retention rate can reach 95% in recycled fibers and was much higher than in other high quality fibers, due to smaller fiber size and larger numbers of fiber fines for recycled fibers [12]. It states again that BC fiber aggregation is a key factor preventing its effective reinforcement to paper made of recycled fibers.

In this study, natural or modified polysaccharides such as xylan, carboxymethyl cellulose, glucomannan, and polyethylene oxide were used as additives in order to improve the dispersion of BC fibers through steric repulsions. BC fibers were also oxidized through TEMPO-mediated oxidation in order to generate negative surface charges so that BC fibers dispersion would be improved through electrostatic repulsion. At the last, the dispersed BC fibers were used to reinforce paper produced from recycled fibers and mechanisms involved in various methods for dispersing BC fibers were also investigated.

2. Materials and Methods

2.1. Materials

Recycled fiber pulp (RFP), made from old newspaper (ONG) and old magazine (OMG), was obtained from Guangzhou Paper Co., Ltd. (Guangzhou, China) with a Canadian Standard Freeness (CSF) of 270 mL; the pulps were kept at moisture content of 90% at 4 °C and were used within 30 days of preparation. CMC-I (Mw = 90,000, DS = 0.7), CMC-II (Mw = 250,000, DS = 1.2), CMC-III (Mw = 250,000, DS = 0.7), nonionic polyethylene oxide (PEO, average Mw ~4,000,000) and sodium hypochlorite solution (6–14% active chlorine basis) were purchased from Shanghai Macklin Biochemical Co., Ltd. (Shanghai, China). Xylan (from sugarcane bagasse, Mw = 30,000) was purchased from Shanghai Yuanye Biochemical Co., Ltd. (Shanghai, China). Glucomannan (from konjac) was purchased from Hubei Yizhi Konjac Biotechnology Co., Ltd. (Hubei, China). Cationized starch (DS = 0.025), which was produced from (3-chloro-2-hydroxypropyl)trimethylammonium chloride modified corn starch, were provided by Guangzhou Paper Co., Ltd. (Guangzhou, China). All other chemicals used are of analytical grade.

2.2. Preparation of Bacterial Cellulose (BC)

Gluconacetobacter xylinus ATCC23767 was obtained from Nanjing High Tech University Biological Technology Research Institute Co., Ltd. (Nanjing, China) and used to produce the bacterial cellulose (BC) pellicles. Static fermentation method was used to produce the BC pellicles. Detailed preparation methods and characterizations of the BC pellicles were shown in Xiang et al. [12,13].

2.3. BC Fiber Dispersion Evaluation

Stock solutions (2 g/L) of CMC-I, xylan, cationic starch, glucomannan, and nonionic PEO were prepared. Approximately 5 g (corresponding to 0.075 g dry weight) of wet BC pellicles (MC = 98.5%) or oxidized-BC were cut into small pieces and placed into a lab blender (SKG-1246, Foshan, China). Given amounts (corresponding to give the weight ratio of macromolecules/BC of 0, 0.05, 0.1, 0.25, and 0.5) of the macromolecule stock solutions were added. Deionized (DI) water was then added to make the mixture volume to 250 mL. The mixture was blended by the lab blender for 2 min to prepare the BC fiber suspension with a BC fiber concentration of 0.3 g/L (dry matter). The BC fiber suspension was poured immediately into a 250 mL-cylinder. The freshly prepared suspensions with BC fiber well dispersed was white and non-transparent in the 250 mL-cylinder. With time, the aggregation of BC fibers resulted in the shortening of the non-transparent portion, which were recorded with time.

For zeta potential measurement, particle concentration in the suspension should be kept between 10^{-5} and 10^{-2} volume fraction [27]. The freshly prepared BC fiber suspensions were diluted 10 times with DI water to a BC fiber concentration of 0.03 g/L (dry matter). The pH value was ~7.0 for all the solutions. Approximately 1.5 mL of the diluted fiber suspensions were then poured into the sample cell and the zeta potential was measured through a nanoparticle analyzer (Horiba SZ-100Z, Kyoto, Japan). The dynamic (absolute) viscosities of the solutions of different dispersants were determined at 25 °C by using a rotational viscometer (Brookfield DV-II HE, Middleboro, MA, USA) at a rotational speed of 250 rpm; the solution concentration was set to 0.15 g/L (dry matter), which was the same as the concentration of the dispersants in BC fiber suspensions when weight ratio of macromolecules/BC was 0.5.

2.4. Adsorption of Macromolecules onto BC Fibers

Stock solutions of CMC-I, xylan, cationic starch, glucomannan, and nonionic PEO with various concentrations (200 mg/L, 600 mg/L, and 1000 mg/L) were prepared. Approximately 1 g of wet BC pellicle (MC = 98.5%) was disintegrated in 250 mL DI water by the lab blender in instant mode three times. The disintegrated BC and 30 mL of the stock solution was added into a conical flask. The mixture has a pH of ~7.0 and was stirred at 350 rpm for 2 h. The BC was then filtered and washed through a sand-cored funnel (pore size 30–50 μm). The filtrate was oven dried and the increment of weight was the macromolecules that did not adsorbed on BC.

2.5. The TEMPO-Mediated Oxidation of BC Fibers

The TEMPO-mediated oxidation of BC fibers was conducted according to Saito et al. [28] with minor modifications. Approximately 20 g of wet BC pellicle (MC = 98.5%) was disintegrated by the lab blender in water in instant mode for three times and the water was filtered out through a sand-cored funnel (pore size 30–50 μm). The disintegrated BC fibers were suspended in water (30 mL) containing TEMPO (0.008 g, 0.05 mmol) and sodium bromide (0.05 g, 0.5 mmol). Into the BC fiber suspension, 0.5 mL, 1 mL, and 2 mL of the NaClO solution (6–14% active chlorine basis) were added. The pH was adjusted 10 by the addition of 0.1 M HCl. The TEMPO-mediated oxidation was continued at room temperature by stirring at 350 rpm for 3 h. During the reaction, the pH was maintained at 10 by adding 0.5 M NaOH. The oxidized BC fibers were filtered through a sand-cored funnel (pore size 30–50 μm) and washed with DI water until pH neutral. The samples were denoted OBC-0.5, OBC-1, and OBC-2, respectively, according to the volume of NaClO solution added to the oxidation process, e.g., 0.5 mL, 1 mL, and 2 mL.

2.6. Paper Handsheet Preparation and Characterization

BC membranes were mixed with stock solutions of CMC, xylan, cationic starch, glucomannan, or nonionic PEO and mechanically disintegrated by a lab blender (SKG-1246, Foshan, China) for 5 min. The recycled fiber pulps and the disintegrated BC fibers were mixed with 1% ratio (proportion of BC based on the total fiber dry weight) and dispersed with a standard pulp-disintegrator at a consistency of 1% (*m/m*) for 15,000 r. The hand sheets were made from the mixed pulp through a standard sheet former (Labtech 200-1, Laval, QC, Canada). The grammage of each sheet was maintained at 70 g/m^2 (dry weight) level. The sheets were dried and equilibrated for 24 h at 23 °C and 50% humidity prior to physical tests. The dry tensile index and air permeability of the sheets were tested according to TAPPI standards (T494 and T251 wd-96, respectively). For wet tensile strength measurement, the middle part of the specimens was soaked in water for 8 s and the specimens were tested immediately for the tensile index. Paper sheet surface morphologies were evaluated using SEM (Zeiss Evo-18, Munich, Germany).

3. Results and Discussion

3.1. BC Fiber Dispersion and Reinforcement of Paper Handsheets

Our previous studies have already proved that one of the key factors limiting the reinforcement of paper by BC fibers is the aggregation of BC fibers [12]. It was hypothesized that if BC fibers were well dispersed during sheet formation, the strength of paper sheet would be effectively improved. Therefore, to evaluate if BC fiber dispersion has positive effects on BC reinforced paper, the BC fibers dispersed by various macromolecules were added to pulps of recycled fibers to make paper sheets. The macromolecules used are CMC-I (Mw = 90,000, DS = 0.7), CMC-II (Mw = 250,000, DS = 1.2), CMC-III (Mw = 250,000, DS = 0.7), nonionic polyethylene oxide (PEO, average Mw = 4,000,000), xylan (from sugarcane bagasse, Mw = 30,000), glucomannan (from konjac), and cationized starch (DS = 0.025). The air permeability, dry tensile strength, and wet tensile strength and air permeability of BC-reinforced paper were investigated (Figure 1). Air permeability of the paper reinforced by BC may reflect the dispersion of BC to some extent. Lower air permeability indicated better BC dispersion, as found by previous study of reinforcing paper with nano-cellulose [3]. From Figure 1a, after adding dispersed BC, the air permeability of paper decreased, indicating a good dispersion of BC by macromolecules. Glucomannan and cationized starch dispersed BC gave the lowest air permeability to paper, which may correspond to their great dispersing effects to BC. From Figure 1b, Being dispersed by most of the macromolecules, the BC reinforced paper had improved dry tensile index, from 32.7 N·m/g to 37.7 N·m/g when adding CMC-I, to ~37 N·m/g when adding xylan or glucomannan, and to 40.7 N·m/g when adding cationized starch. The percentage improvement was from 15% to 25%. Without adding BC fibers, some macromolecules themselves can already improve the paper strength, such as CMC-I and cationized starch, but adding BC can further improve the tensile strength of paper. By adding only cationized starch, the paper strength had already improved to 37.2 N·m/g. Cationized starch helps to improve the retention of BC fibers or fines from recycled fibers and thus improves the paper strength [29]. The best improvement in paper strength caused by macromolecule dispersed BC was obtained with glucomannan by which an improvement of 4.2 N·m/g or 12.7% was obtained. This may correlate to that glucomannan have the best dispersing ability as demonstrated by the air permeability data (Figure 1a). From Figure 1c, the wet tensile index of paper sheet had not much improvement after adding dispersed BC, because BC fibers and recycled fibers connect through hydrogen bonding that is weak in wet. Adding only CMC improved the wet tensile index of paper a lot, which may be due to the forming of covalent bonds between CMC and plant fibers [30]. After adding BC, CMC tended to bind with BC resulting in the decrease of paper wet tensile strength.

In sum, it can be concluded here that improved dispersion of BC fiber is helpful to improve its reinforcement to paper dry tensile strength. However, the improvement was not very high with up to ~12% caused by macromolecule-dispersed BC (taking out the reinforcement effects of macromolecules only). This showed that the quality of pulp fibers is still very important to the reinforcement effects of BC fibers. Proper dispersion of BC fibers can compensate for the pulp quality to some extent.

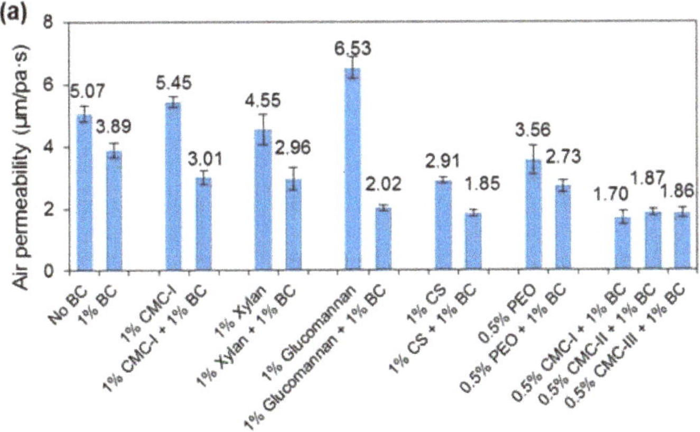

Figure 1. *Cont.*

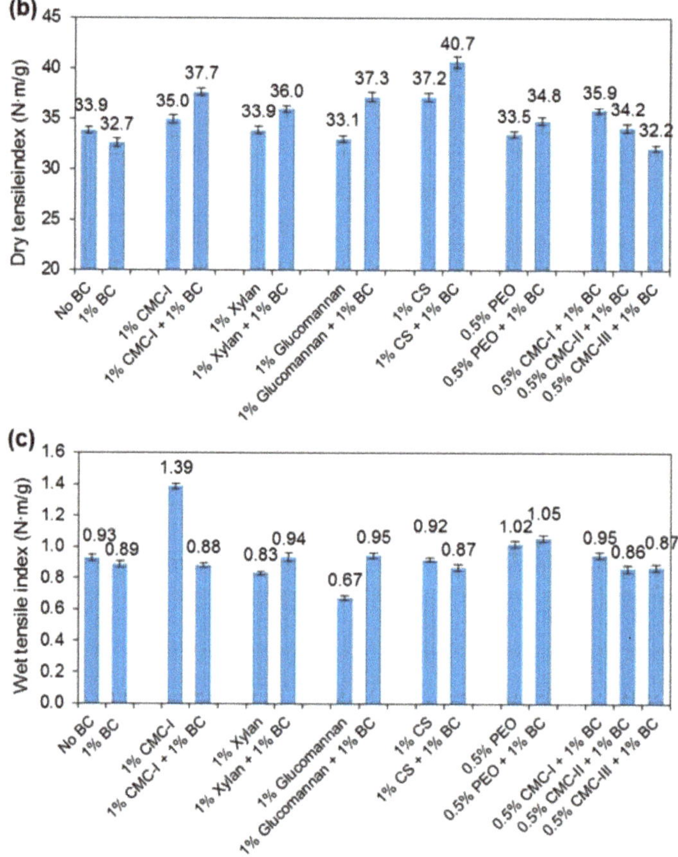

Figure 1. (**a**) Air permeability, (**b**) dry tensile index, and (**c**) wet tensile index of paper hand sheets produced from recycled fibers and different additive macromolecules.

3.2. Effects of Additives on BC Fiber Dispersion

As discussed in Section 3.1, once dispersed by various macromolecules, the BC fibers improved tensile strength of paper sheets significantly (Figure 1). However, the dispersing effects of various macromolecules on BC fiber suspension should be evaluated and correlated with paper sheet tensile strength in order to understand the mechanism for reinforcement of paper by dispersed BC. Therefore, carboxymethyl cellulose (CMC), xylan, glucomannan, cationized starch, and polyethylene oxide (PEO) were added during the disintegration of BC fibers. The total volume of BC fiber suspension was made to 250 mL (0.3 g/L) and kept in a 250 mL-cylinder. The changes of the height of BC fiber fractions with time were measured as indicator for the dispersing effects of different macromolecules (Figure 2). Adhesion of macromolecules to cellulose fiber surface may also improve the fiber dispersion through steric hindrance effects [19]. If the suspension is less stable, the BC fibers aggregate faster and thus the height of the BC fiber fraction decreases faster. From Figure 2a–e, the macromolecule leading the best colloidal stability of BC fiber suspension was glucomannan, with which the suspension had no aggregation in 50 min at 0.1 weight ratio of addition. The BC fiber suspensions with cationized starch and xylan had no aggregation of BC fiber at 0.25 weight ratio of addition, while suspensions with CMC-I and PEO had no aggregation at 0.5 weight ratio of addition.

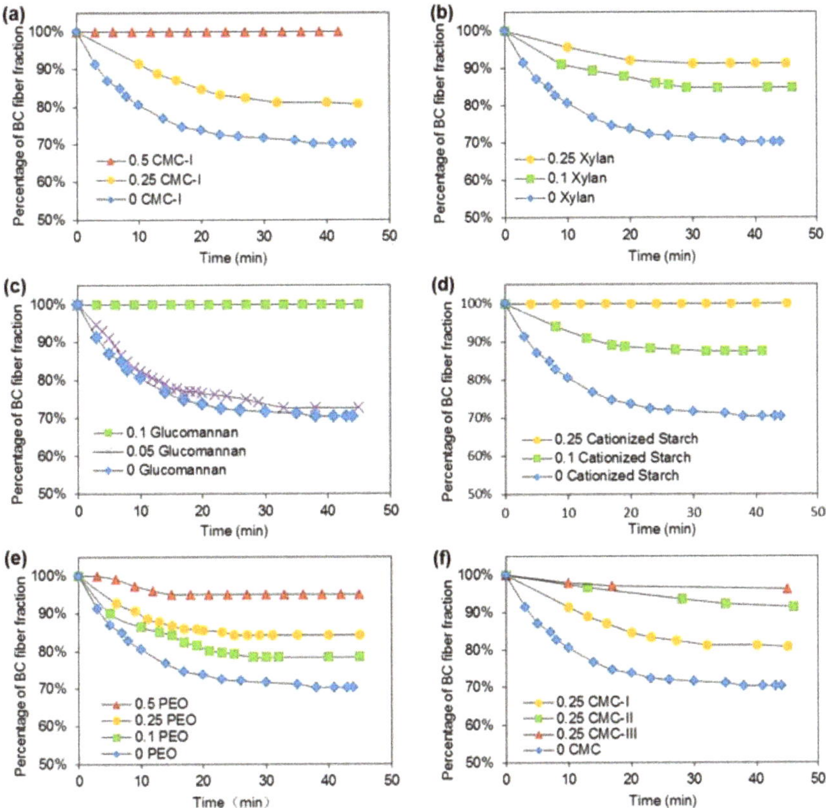

Figure 2. The changes of the percentage height of BC fiber fractions in the 250 mL-cylinder with time for BC fiber suspensions dispersed by (**a**) CMC-I, (**b**) xylan, (**c**) glucomannan, (**d**) cationized starch, (**e**) PEO, and (**f**) CMC.

CMC-I has Mw = 90,000 and DS = 0.7. To evaluate the effects of molecular weight and viscosity of the macromolecules on the dispersion of BC fibers, CMC-II (Mw = 250,000, DS = 1.2) and CMC-III (Mw = 250,000, DS = 0.7) were also used to disperse BC fibers (Figure 2f). At 0.25 weight ratio of addition, CMC-III can prevent BC fibers from aggregation in 50 min, while CMC-I and -II dispersed BC fiber still had some aggregation. Therefore, the dispersing ability of CMC has an order of CMC-III > CMC-II > CMC-I. CMC-III had the same molecular weight but higher viscosity than CMC-II, and both CMC-III and -II had higher molecular weight and viscosity than CMC-I. This result postulated that, for the same macromolecule, molecular weight has positive effects on its dispersing ability for BC fibers, which could be due to the fact that high molecular weight may induce high steric hindrance. In the case of CMC with a same molecular weight, CMC-III has a lower DS than CMC-II and, thus, has more available hydroxyls. High numbers of hydroxyls induce a strong interaction with water resulting in high viscosity of CMC-III [31]. A strong interaction between CMC-III and water may result in its better dispersing effects to BC, compared to CMC-II.

To further investigate the stabilization mechanism of BC fiber suspensions by different macromolecules, viscosity of the macromolecules (Table 1), zeta potential of BC fiber suspension (Table 2), and adsorption of macromolecules on BC fibers (Figure 3) were evaluated. The viscosities of the macromolecules were measured at 0.15 g/L, which was the same as their concentrations in BC fiber suspensions of 0.5 weight ratios. The concentration of the macromolecules in BC fiber suspension was low, resulting in very low viscosities for all the macromolecule solutions (Table 1). This indicated that viscosity was not a major factor affecting BC fiber dispersion in this case. For the zeta potential, the plain BC fiber suspension without any macromolecules added had a value of −21.2 (Table 2). By increasing the weight ratio of macromolecules to BC fibers from 0.1–0.5, the zeta potential of the BC fiber suspension with CMC-I changed from −34.2 to −60.6, with xylan changing from −21.9 to −24.3, with cationized starch from −20.0 to 29.9, and with PEO from −26.6 to −10.3 (Table 2). A high absolute value of the zeta potential usually indicates a high colloidal stability of a suspension system. However, the 0.25 weight ratio of cationized starch dispersed BC fibers had lower absolute value of zeta potential than that of 0.1 weight ratio, but had better dispersion, as shown in Figure 2; BC fiber suspension with 0.25 CMC-I had the highest zeta potential, but did not have the best colloidal stability. These phenomena indicated that there were other factors affecting the colloidal stability. Consequently, the adsorption capacity of macromolecules onto BC fibers (Figure 3) was evaluated. The adsorptions of macromolecules onto fibers were believed to have an important influence of the dispersing of cellulose fibers, since the macromolecules adsorbed by fibers can increase the steric hindrance [19]. From Figure 3, it can be seen that, at high concentration, glucomannan had the highest adsorption to BC fibers, while the adsorptions of cationized and xylan were also high. The adsorptions of CMC and PEO to BC fibers were the lowest and much smaller than other macromolecules. As shown in Figure 2, glucomannan had the best dispersing ability for BC fibers, while cationized starch and xylan were the second best. SEM images showed that glucomannan dispersed BC fibers were more evenly distributed in the paper fiber matrix compared to undispersed BC fibers (Figure 4). From Figure 3, it can be postulated that, the adsorption ability of the macromolecules onto BC fibers was very important to their dispersing effects. In another words, steric hindrance is a major factor for BC fiber dispersion.

Table 1. Viscosity (mean ± STDEV) of different macromolecules at 0.15 g/L concentration and measured at 250 rpm rotational speed of the rotational viscometer.

Additives	CMC-I	Xylan	Glucomannan	Cationized Starch	PEO
Viscosity (mPa·s)	30.9 ± 1.9	28.9 ± 0.6	29.4 ± 1.3	29.9 ± 0.8	33.1 ± 2.3

Table 2. Zeta potential (mean ± STDEV) of BC fiber suspensions with additions of different macromolecules.

Weight Ratio	Zeta potential (mV) of Macromolecule Additives					
	None	CMC-I	Xylan	Glucomannan	Cationized Starch	PEO
0.1	−21.2 ± 0.6	−34.2 ± 1.8	−22.9 ± 0.9	−23.5 ± 1.3	−20.0 ± 0.8	−26.6 ± 2.0
0.25	-	−56.8 ± 2.1	−24.9 ± 1.1	-	−13.2 ± 0.4	−13.2 ± 1.4
0.5	-	−60.6 ± 1.7	−24.3 ± 1.0	-	29.9 ± 2.2	−10.3 ± 0.7

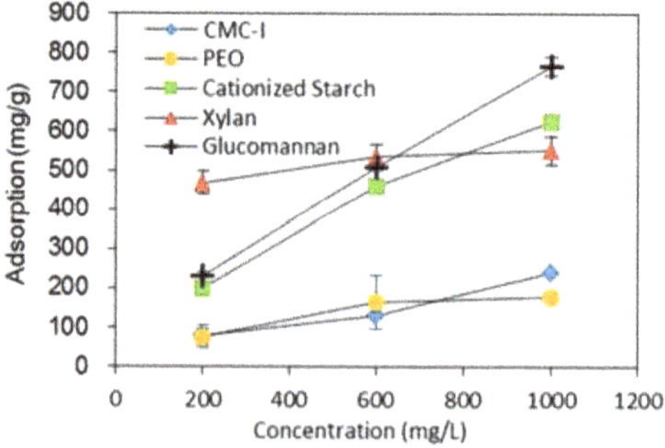

Figure 3. The adsorption of different macromolecules onto BC fibers at different concentrations.

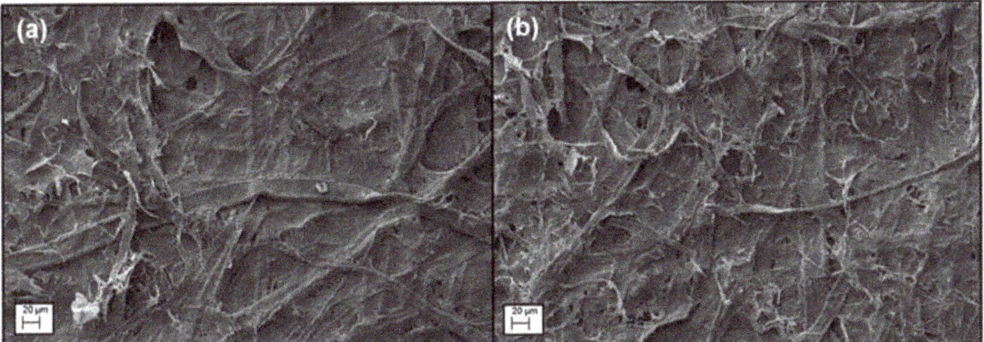

Figure 4. SEM images of recycled fiber paper reinforced by (**a**) undispersed BC and (**b**) glucomannan dispersed BC.

3.3. BC Fiber Dispersion by Surface Oxidation

Electrostatic repulsion is another important mechanism to improve BC fiber dispersion. TEMPO-mediated oxidation method was used to oxidize the surface of BC fibers. The C6 on anhydroglucose unit is selectively oxidized into carboxylic acid [28] and the negative charges on BC fiber surfaces may repulse each other preventing the fiber from aggregation. The oxidation was controlled at a low level avoiding the excessive decrease of the hydrogen bonds between BC fibers and plant fibers, as well as preventing the degradation of BC. BC fiber samples with different oxidized levels were numbered as OBC-2, OBC-1, and OBC-0.5, and their carboxyl contents were measured as 1.78, 1.38, and 0.41 mmol/g, respectively.

The suspensions (250 mL and 0.3 g/L) produced from all of the three oxidized BC samples did not aggregate within 50 min, indicating the TEMPO-mediated oxidation effectively improves the BC fiber dispersion or colloidal stability. The oxidized BC fibers were mixed with recycled fibers to produce reinforced paper. Figure 5a showed that the air permeability of the recycled fiber decreased by adding OBC, indicating OBC was well dispersed during paper forming [3]. At high oxidation level, the OBC reinforced paper had slightly increased air permeability may be due to the excessive degradation of BC. Figure 5b showed that adding TEMPO-mediated oxidized BC fiber does not help to improve the dry tensile index of paper. However, the wet tensile index improved greatly, especially for the BC fiber with the lowest oxidation level (OBC-0.5), i.e., from 0.89 N·m/g to 1.59 N·m/g, increased by 80% (Figure 5c). The increase in wet tensile strength indicated covalent bonds formed between fibers [32]. The explanation was that some hydroxyls were not oxidized into carboxyls, but into aldehydes [28,33], especially at low oxidation level. The aldehydes may form hemiacetals or acetals with fiber hydroxyls, leading to the improvement in paper wet strength [32]. However, the numbers of aldehyde were low and resulted in no increase in dry tensile strength, since the values of dry tensile index were high.

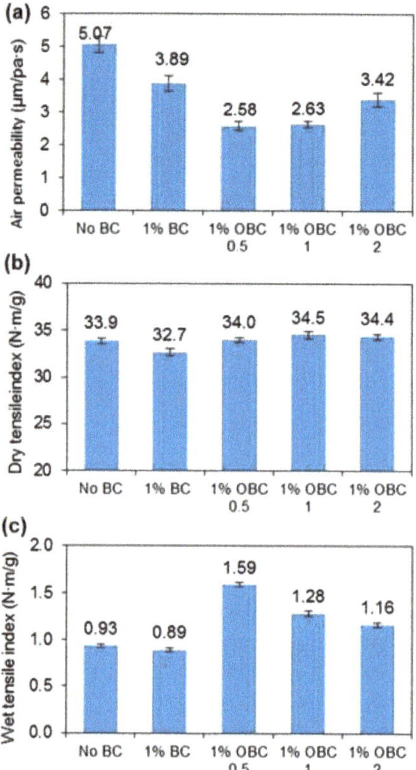

Figure 5. (**a**) Air permeability, (**b**) dry tensile index, and (**c**) wet tensile index of paper hand sheets produced from recycled fibers and oxidized BC fibers.

The oxidized BC fibers leading to limited paper strength improvement may be due to the competition of hydrogen bonds and carboxyl groups from the oxidation process. BC fibers reinforce paper by increasing hydroxyl hydrogen bonding sites between plant fibers. The carboxyl groups are introduced by the TEMPO-mediated oxidation process, and the more of the carboxyls, the better is the BC dispersion. However, carboxyls are transformed from hydroxyls. Therefore, the transformation of carboxyl groups causes the decrease of the number of hydrogen bonds between BC fibers and plant fibers, limiting the improvement of paper strength. Another major reason for limited paper strength improvement may be due to the TEMPO-mediated oxidation process degraded BC to some extent.

In conclusion of this section, TEMPO-mediated oxidation is effective in improving the colloidal stability of BC fiber suspensions, but not effective in improving the BC fiber ability to enhance paper dry tensile index although the paper wet tensile index was improved.

4. Conclusions

In this study, natural or modified polysaccharides, such as xylan, CMC, glucomannan, and cationized starch were used as additives to improve the dispersion of BC fibers or its reinforcing effects to recycled fiber paper. Good dispersion of BC fiber is helpful to its reinforcement to the dry tensile strength of paper made from recycled fiber, even though the improvement was not very high, with up to ~12% caused by macromolecule-dispersed BC. This showed that the quality of pulp fibers is still very important to the reinforcement effects of BC fibers. Proper dispersion of BC fibers can compensate for the pulp quality to some extent.

The adsorption of the macromolecules onto BC fibers is one of the key factors for their ability to disperse BC fibers. Some macromolecules, such as cationized starch, can evidently improve the paper strength alone without adding BC due to their fiber retention ability, but its adsorption to BC fibers can still improve BC dispersion and, thus, further improve the paper strength. The dispersing effects of the macromolecules can be correlated with the tensile strength of paper sheets. Among the macromolecules studied, glucomannan had the best dispersing ability to BC fibers and led to the best paper tensile strength improvement.

TEMPO-mediated oxidation is effective in improving the colloidal stability of BC fiber suspensions, but not effective in improving the BC fiber ability to enhance paper dry tensile index, probably because BC was degraded by the oxidation process. However, the incomplete oxidation of hydroxyls to aldehyde on BC fibers helps to improve the paper wet tensile index.

Author Contributions: Original Idea, Z.X. and F.L.; Methodology, Z.X., Q.L., and Y.C.; BC Production, Q.L. and Y.C.; Data Collection, J.Z.; Data Curation, Z.X. and J.Z.; Writing-Original Draft Preparation, Z.X.; Writing-Review & Editing, J.L. and F.L.; Project Administration, J.L. and F.L.; Funding Acquisition, Z.X. and F.L.

Funding: This research was funded by the Guangzhou Science and Technology Program (Key Scientific Research Project 201707020011), the National Natural Science Foundation of China (31600470), the Guangzhou Science and Technology Program (General Scientific Research Project 201707010053), and the Guangdong Province Science Foundation for Cultivating National Engineering Research Center for Efficient Utilization of Plant Fibers (2017B090903003). The APC was funded by the Guangzhou Science and Technology Program (Key Scientific Research Project 201707020011).

Conflicts of Interest: The authors declare no conflict of interest.

Abbreviations

BC	Bacterial cellulose
CMC	Carboxymethyl cellulose
OBC	Oxidized bacterial cellulose
PEO	Polyethylene oxide

References

1. Japan Paper Association. Available online: http://www.jpa.gr.jp/states/global-view/index.html (accessed on 16 October 2018).
2. Araki, J. Electrostatic or steric?—Preparations and characterizations of well-dispersed systems containing rod-like nanowhiskers of crystalline polysaccharides. *Soft Matter* **2013**, *9*, 4125–4141. [CrossRef]
3. Campano, C.; Merayo, N.; Balea, A.; Tarres, Q.; Delgado-Aguilar, M.; Mutje, P.; Negro, C.; Blanco, A. Mechanical and chemical dispersion of nanocelluloses to improve their reinforcing effect on recycled paper. *Cellulose* **2018**, *25*, 269–280. [CrossRef]
4. Kose, R.; Yamaguchi, K.; Okayama, T. Preparation of fine fiber sheets from recycled pulp fibers using aqueous counter collision. *Cellulose* **2016**, *23*, 1393–1399. [CrossRef]
5. Castro, D.O.; Karim, Z.; Medina, L.; Häggström, J.O.; Carosio, F.; Svedberg, A.; Wågberg, L.; Söderberg, D.; Berglund, L.A. The use of a pilot-scale continuous paper process for fire retardant cellulose-kaolinite nanocomposites. *Compos. Sci. Technol.* **2018**, *162*, 215–224. [CrossRef]
6. Balea, A.; Merayo, N.; Fuente, E.; Delgado-Aguilar, M.; Mutje, P.; Blanco, A.; Negro, C. Valorization of Corn Stalk by the Production of Cellulose Nanofibers to Improve Recycled Paper Properties. *Bioresources* **2016**, *11*, 3416–3431. [CrossRef]
7. Balea, A.; Merayo, N.; Fuente, E.; Negro, C.; Delgado-Aguilar, M.; Mutje, P.; Blanco, A. Cellulose nanofibers from residues to improve linting and mechanical properties of recycled paper. *Cellulose* **2018**, *25*, 1339–1351. [CrossRef]
8. Merayo, N.; Balea, A.; de la Fuente, E.; Blanco, A.; Negro, C. Synergies between cellulose nanofibers and retention additives to improve recycled paper properties and the drainage process. *Cellulose* **2017**, *24*, 2987–3000. [CrossRef]
9. Jonas, R.; Farah, L.F. Production and application of microbial cellulose. *Polym. Degrad. Stabil.* **1998**, *59*, 101–106. [CrossRef]

10. Iguchi, M.; Yamanaka, S.; Budhiono, A. Bacterial cellulose—A masterpiece of nature's arts. *J. Mater. Sci.* **2000**, *35*, 261–270. [CrossRef]
11. Yoshinaga, F.; Tonouchi, N.; Watanabe, K. Research progress in production of bacterial cellulose by aeration and agitation culture and its application as a new industrial material. *Biosci. Biotechnol. Biochem.* **1997**, *61*, 219–224. [CrossRef]
12. Xiang, Z.; Jin, X.; Liu, Q.; Chen, Y.; Li, J.; Lu, F. The reinforcement mechanism of bacterial cellulose on paper made from woody and non-woody fiber sources. *Cellulose* **2017**, *24*, 5147–5156. [CrossRef]
13. Xiang, Z.; Liu, Q.; Chen, Y.; Lu, F. Effects of physical and chemical structures of bacterial cellulose on its enhancement to paper physical properties. *Cellulose* **2017**, *24*, 3513–3523. [CrossRef]
14. Xiang, Z.; Chen, Y.; Liu, Q.; Lu, F. A highly recyclable dip-catalyst produced from palladium nanoparticle-embedded bacterial cellulose and plant fibers. *Green Chem.* **2018**, *20*, 1085–1094. [CrossRef]
15. Jin, X.; Xiang, Z.; Liu, Q.; Chen, Y.; Lu, F. Polyethyleneimine-bacterial cellulose bioadsorbent for effective removal of copper and lead ions from aqueous solution. *Bioresour. Technol.* **2017**, *244*, 844–849. [CrossRef] [PubMed]
16. Amornkitbamrung, L.; Marnul, M.-C.; Palani, T.; Hribernik, S.; Kovalcik, A.; Kargl, R.; Stana-Kleinschek, K.; Mohan, T. Strengthening of paper by treatment with a suspension of alkaline nanoparticles stabilized by trimethylsilyl cellulose. *Nano-Struct. Nano-Objects* **2018**, *16*, 363–370. [CrossRef]
17. Cavallaro, G.; Danilushkina, A.A.; Evtugyn, V.G.; Lazzara, G.; Milioto, S.; Parisi, F.; Rozhina, E.V.; Fakhrullin, R.F. Halloysite Nanotubes: Controlled Access and Release by Smart Gates. *Nanomaterials* **2017**, *7*, 199. [CrossRef] [PubMed]
18. Cavallaro, G.; Lazzara, G.; Milioto, S.; Parisi, F. Halloysite Nanotubes for Cleaning, Consolidation and Protection. *Chem. Rec.* **2018**, *18*, 940–949. [CrossRef]
19. Winter, H.T.; Cerclier, C.; Delorme, N.; Bizot, H.; Quemener, B.; Cathala, B. Improved Colloidal Stability of Bacterial Cellulose Nanocrystal Suspensions for the Elaboration of Spin-Coated Cellulose-Based Model Surfaces. *Biomacromolecules* **2010**, *11*, 3144–3151. [CrossRef]
20. Mishima, T.; Hisamatsu, M.; York, W.S.; Teranishi, K.; Yamada, T. Adhesion of β-D-glucans to cellulose. *Carbohydr. Res.* **1998**, *308*, 389–395. [CrossRef]
21. Pirich, C.L.; de Freitas, R.A.; Woehl, M.A.; Picheth, G.F.; Petri, D.F.S.; Sierakowski, M.R. Bacterial cellulose nanocrystals: Impact of the sulfate content on the interaction with xyloglucan. *Cellulose* **2015**, *22*, 1773–1787. [CrossRef]
22. Azzam, F.; Heux, L.; Putaux, J.L.; Jean, B. Preparation By Grafting Onto, Characterization, and Properties of Thermally Responsive Polymer-Decorated Cellulose Nanocrystals. *Biomacromolecules* **2010**, *11*, 3652–3659. [CrossRef] [PubMed]
23. Kloser, E.; Gray, D.G. Surface Grafting of Cellulose Nanocrystals with Poly(ethylene oxide) in Aqueous Media. *Langmuir* **2010**, *26*, 13450–13456. [CrossRef] [PubMed]
24. Cheng, D.W.Y.; An, X.; Zhu, X.; Cheng, X.; Zheng, L.; Nasrallah, J.E. Improving the colloidal stability of Cellulose nano-crystals by surface chemical grafting with polyacrylic acid. *J. Bioresour. Bioprod.* **2016**, *1*, 114–119.
25. Zaman, M.; Xiao, H.N.; Chibante, F.; Ni, Y.H. Synthesis and characterization of cationically modified nanocrystalline cellulose. *Carbohydr. Polym.* **2012**, *89*, 163–170. [CrossRef] [PubMed]
26. Hasani, M.; Cranston, E.D.; Westman, G.; Gray, D.G. Cationic surface functionalization of cellulose nanocrystals. *Soft Matter* **2008**, *4*, 2238–2244. [CrossRef]
27. Sun, C.H.; Gunasekaran, S. Effects of protein concentration and oil-phase volume fraction on the stability and rheology of menhaden oil-in-water emulsions stabilized by whey protein isolate with xanthan gum. *Food Hydrocoll.* **2009**, *23*, 165–174. [CrossRef]
28. Saito, T.; Kimura, S.; Nishiyama, Y.; Isogai, A. Cellulose nanofibers prepared by TEMPO-mediated oxidation of native cellulose. *Biomacromolecules* **2007**, *8*, 2485–2491. [CrossRef]
29. Tovar, R.G.; Fischer, W.J.; Eckhart, R.; Bauer, W. White Water Recirculation Method as a Means to Evaluate the Influence of Fines on the Properties of Handsheets. *Bioresources* **2015**, *10*, 7242–7251.
30. Pereira, P.H.F.; Waldron, K.W.; Wilson, D.R.; Cunha, A.P.; de Brito, E.S.; Rodrigues, T.H.S.; Rosa, M.F.; Azeredo, H.M.C. Wheat straw hemicelluloses added with cellulose nanocrystals and citric acid. Effect on film physical properties. *Carbohydr. Polym.* **2017**, *164*, 317–324. [CrossRef]

31. Xiang, Z.Y.; Runge, T. Emulsifying properties of succinylated arabinoxylan-protein gum produced from corn ethanol residuals. *Food Hydrocoll.* **2016**, *52*, 423–430. [CrossRef]
32. Xiang, Z.Y.; Anthony, R.; Lan, W.; Runge, T. Glutaraldehyde crosslinking of arabinoxylan produced from corn ethanol residuals. *Cellulose* **2016**, *23*, 307–321. [CrossRef]
33. Isogai, A.; Saito, T.; Fukuzumi, H. TEMPO-oxidized cellulose nanofibers. *Nanoscale* **2011**, *3*, 71–85. [CrossRef] [PubMed]

 *nanomaterials*

Article

Synthesis of Nanofibrillated Cellulose by Combined Ammonium Persulphate Treatment with Ultrasound and Mechanical Processing

Inese Filipova [1,*], Velta Fridrihsone [1], Ugis Cabulis [1] and Agris Berzins [2]

[1]   Latvian State Institute of Wood Chemistry, Dzerbenes Street 27, LV-2121 Riga, Latvia;
       fridrihsone.velta@inbox.lv (V.F.); cabulis@edi.lv (U.C.)

[2]   Faculty of Chemistry, University of Latvia, Jelgavas Street 1, LV-1004 Riga, Latvia; agris.berzins@lu.lv

*   Correspondence: inese.filipova@inbox.lv; Tel.: +371-291-418-82

Received: 30 June 2018; Accepted: 16 August 2018; Published: 21 August 2018

Abstract: Ammonium persulfate has been known as an agent for obtaining nanocellulose in recent years, however most research has focused on producing cellulose nanocrystals. A lack of research about combined ammonium persulfate oxidation and common mechanical treatment in order to obtain cellulose nanofibrils has been identified. The objective of this research was to obtain and investigate carboxylated cellulose nanofibrils produced by ammonium persulfate oxidation combined with ultrasonic and mechanical treatment. Light microscopy, atomic force microscopy (AFM), powder X-Ray diffraction (PXRD), Fourier-transform infrared spectroscopy (FTIR), thermogravimetric analysis (TGA) and Zeta potential measurements were applied during this research. The carboxylated cellulose suspension of different fractions including nanofibrils, microfibrils and bundles were produced from bleached birch Kraft pulp fibers using chemical pretreatment with ammonium persulfate solution and further defibrillation using consequent mechanical treatment in a high shear laboratory mixer and ultrasonication. The characteristics of the obtained nanofibrils were: diameter 20–300 nm, crystallinity index 74.3%, Zeta potential -26.9 ± 1.8 mV, clear FTIR peak at 1740 cm^{-1} indicating the C=O stretching vibrations, and lower thermostability in comparison to the Kraft pulp was observed. The proposed method can be used to produce cellulose nanofibrils with defined crystallinity.

Keywords: nanocellulose; ammonium persulfate; oxidation; nanofibrils; high shear mixer

1. Introduction

Cellulose is a natural, linear, renewable biopolymer, composed of D-glucopyranose units, and it is present naturally in all plants on earth. Nanocellulose is material obtained by the disintegration of cellulose into nanoscale particles such as cellulose nanowhiskers (CNC), cellulose nanofibers (CNF), cellulose nanospheres (CNS) and amorphous nanocellulose (ANC) [1]. The above-mentioned nanoscale particles are extensively investigated for a wide range of different uses in eco-friendly advanced applications and materials, such as nanocomposites, bionanomaterials and others [2–6]. The shape of the cellulose nanoscale particles depends both on the source [7] and method of production [8]. CNC are obtained mainly by acid hydrolysis of cellulose fibers [9], whereas CNF are usually extracted by mechanical disintegration of cellulose fiber combined with biological and/or chemical pre-treatment [10]. CNF, also known as nanofibrillated cellulose, microfibrillated cellulose (MFC) and cellulose nanofibrils, is involved in a large number of advanced applications, such as nanocomposites [11,12], foams, aerogels [13], packaging [14] and others.

CNF have a high surface area covered with hydroxyl groups, which provide options for surface chemical modification to differentiate between the properties and characteristics of the targe material. A large number of modification methods have been investigated in recent years, such as esterification [15], etherification, silylation, urethanization, amidation [16], click reactions [17], and polymer grafting [18]. Oxidation is known as the most commonly used method to convert cellulose into value-added derivatives [19]. Isogai et al. [20] suggested 2,2,6,6,-tetramethylpiperidine-1-oxyl radical (TEMPO)-oxidation as a simultaneous process for CNF production and its functionalization. The above-mentioned method is now widely used as a pre-treatment in order to produce carboxylated CNF for the subsequent use in a wide range of applications [21–25]. The disadvantage of this method is that TEMPO is very expensive and toxic.

Ammonium persulfate (APS) is known as a strong oxidizer with low long-term toxicity, high water solubility and low cost [26]. The persulfate ion $S_2O_8^{2-}$ can be activated using heat or light, and the resulting $SO_4\bullet^-$ initiates a chain of reactions involving other radicals and oxidants [27]. The first paper about the use of APS for the production of nanocellulose was published by Leung et al. [26], where a one-step procedure for producing high crystalline carboxylated CNC from different cellulosic materials was described. Currently, APS oxidation is mentioned as one of the promising oxidation methods to produce CNC [28]. APS oxidation has the potential to be used for carboxylated CNC production as an alternative to TEMPO-mediated oxidation [29]. The main advantage of the APS method is that there is no need for purification of the raw material before nanocellulose production. Since 2011, the APS method has become widely used and a number of articles about obtaining of CNC by the APS method from different feedstock, such as cotton linter [26,30–33], flax [26,34], hemp [26,34], triticate [26], micro crystalline cellulose [26,33–36], wood pulp [26,30,32,37,38], bacterial cellulose [26,39,40], borer powder of bamboo [41], cornhob [42], Lyocell [43], oil palm empty fruit bunch [44], coconut fiber [45], and sugarcane bagasse [29,46], have been published.

Most of the articles are focused on the production of carboxylated CNC by heating the cellulose materials at 60–90 °C with 1–1.5 M APS solution for 8–24 h. As a result, CNC with a diameter of 3–100 nm and length between 100 and 500 nm are obtained. CNS were obtained from sugarcane bagasse using 2 M APS solution for 16 h at 60 °C [29], from bamboo powder using 2 M APS solution for 6 h at 65 °C [33] and from Lyocell fibers using 1 M APS solution at 70 °C with various oxidation times [43]. It was reported by Goh et al. that MFC was obtained from oil palm empty fruit bunch using APS oxidation combined with ultrasonication [44].

A lack of research about combined APS oxidation and common mechanical treatment has been identified. It was reasonable to put forward a hypothesis about obtaining oxidized CNF by combining APS treatment with ultrasound and rather mild mechanical processing. Therefore, the objective of the current research was to obtain and investigate carboxylated CNF produced by APS oxidation combined with ultrasonic and mechanical treatment.

2. Materials and Methods

2.1. Chemical Treatment of Cellulose

Ten grams (dry weight) of total chlorine free bleached birch Kraft pulp (obtained from Sodra Cell AB, Sweden as dry pulp sheets) was soaked in distilled water (500 mL) for 8 h and then disintegrated to 75,000 revolutions in the Disintegrator (Frank PTI, Laakirchen, Austria). The excess water was drained using a Büchner funnel, and the pulp was repeatedly mixed with fresh distilled water (300 mL) in glass beaker. The beaker was covered in foil and the suspension was heated to 70 °C in a water bath. Afterwards, APS (purity, $\geq$98%; 114 g), HCl (concentration 37%; 20.53 mL) and distilled water were added to reach a final volume of 500 mL. The mixture was then heated at 70 °C for 4 h with continuous stirring. The reaction was stopped by cooling the mixture in an ice bath until it reached ~15 °C, and then oxidized cellulose was filtered using a Büchner funnel and washed until it reached the pH of distilled water.

2.2. Mechanical Treatment of Cellulose

The mechanical treatment of cellulose fibers was performed in a two-step procedure. First, 400 mL of oxidized cellulose slurry (cellulose content 2 wt %) was treated in high shear laboratory mixer, Silverson L5M-A (Silverson Machines, Inc., East Longmeadow, MA, USA), at 10,000 rpm for 20 min. The working head of the mixer was completely immersed in the slurry about 1 cm from the bottom of the beaker. Mechanical treatment was followed by ultrasound treatment by an ultrasonic homogenizer, SONIC-650W (MRC Ltd., Holon, Israel), for 8 min (probe diameter 10 mm, 95% of power, 25 Hz, 9 s on, 1 s off). The beaker with the slurry was cooled in the ice bath during the treatment. The combined processing cycle of the sample was repeated seven times until the viscosity visually increased, indicating the occurrence of nanocellulose.

2.3. Characterisation of Nanocellulose

A drop (0.04 μL) of diluted suspension (0.001 wt %) was placed on a microscope slide and allowed to dry at room temperature. Then, atomic force microscopy (AFM) measurements were performed in the tapping mode using the Park NX10 (Park Systems Corporation, Suwon, Korea). The size of the objects in the AFM pictures were measured using XEI 1.7.6. software (Park Systems Corp., Santa Clara, CA, USA).

A drop (0.5 mL) of diluted suspension (0.05 wt %) was placed on a microscope slide, covered with a coverslip and investigated by a Leica DMLB microscope connected to a Leica DFC490 video camera (both from Leica Microsystems GmbH, Wetzlar, Germany) with a magnification of 200–1000×.

For other analyses, the CNF suspension was freeze-dried and milled in a MM200 ball mill (Retsch, Haan, Germany) for 10 min at a frequency of 30 Hz.

The milled sample was mixed with KBr powder and pressed into a small tablet. From this, FTIR spectrum values were recorded using a Nicolet iS50 spectrometer (Thermo Fisher Scientific, Waltham, MA, USA) at a resolution of 4 cm^{-1}, 32 scans per sample.

The thermal stability of APS CNF was compared with Kraft pulp by a TGA SDTA850e thermal analyzer (Mettler Toledo, Columbus, OH, USA) under nitrogen atmosphere, heating rate of 10 °C/min, and temperature ranging from 25 °C to 800 °C; the test sample was approximately 8–10 mg.

The powder X-ray diffraction (PXRD) patterns were measured on a D8 Advance diffractometer (Bruker, Karlsruhe, Germany) using copper radiation at a wavelength of 1.5418 Å, equipped with a LynxEye position sensitive detector. The tube voltage and current were set to 40 kV and 40 mA. The divergence slit was set at 0.6 mm and the anti-scatter slit was set at 8.0 mm. The diffraction patterns were recorded using a 1.0 s/0.02° scanning speed from 5° to 70° on a 2θ scale. Crystallinity of the cellulose sample was calculated using Rietveld refinement on the obtained PXRD patterns in TOPAS v5 (Bruker, Karlsruhe, Germany). The crystalline phase was modeled using crystallographic data of cellulose Iβ (Cambridge Structural Database (CSD) reference code JINROO01). The amorphous phase was modeled according to the partial or no known crystal structures (PONKCS) [47] approach using artificial phase with the PXRD pattern corresponding to amorphous cellulose (prepared by grinding Avicell crystalline cellulose in ball mills for 90 min in stainless steel 35 mL grinding jars with a 20 mm ball at a frequency of 15 Hz), with the calibration constant determined from a measurement of its mixture with corundum. In Rietveld refinement, the background was modeled with the Chebyshev polynomial function of second degree. For crystalline cellulose lattice parameters and crystallite size, the parameters were refined by additionally modeling the preferred orientation with spherical harmonics of sixth order.

The zeta potential was determined on Zeta Sizer Nano ZS90 (Malvern Panalytical Ltd., Malvern, UK) for 0.05 wt % CNF suspension in distilled water.

3. Results and Discussion

3.1. Microscopy

The synthesis of CNF started with the chemical treatment of bleached birch Kraft pulp fibers with APS solution. Structural damages were identified on the surface of fibers after chemical treatment by APS (Figure 1a). There were visible defects in certain fiber sections leading to splitting of the fiber perpendicular to the longitudinal axis. It is well known that the most accessible regions of the cellulose are disordered amorphous regions [1]; therefore, it is clear that the damages observed in Figure 1 are rather related to the destruction of cellulose polymer chains mostly in amorphous regions. The images of APS treated cellulose demonstrated a mixture of particles with sizes in range from 50 µm to 500 µm. The reaction yield was $80 \pm 2\%$ from initial pulp material.

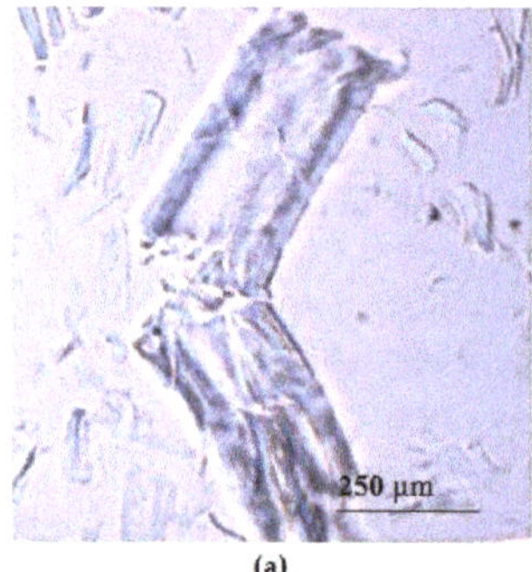
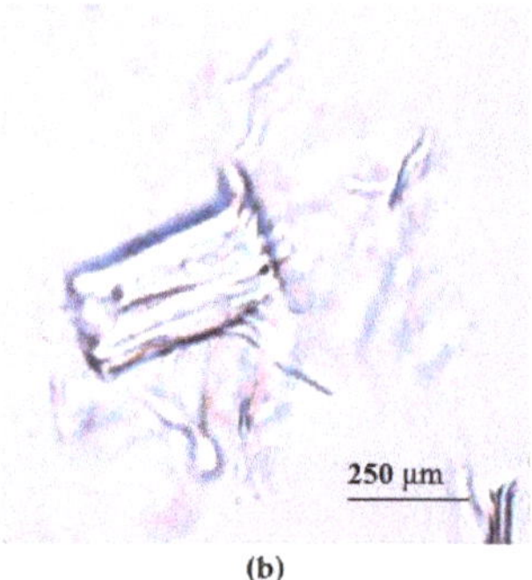

(a)　　　　　　　　　　　　　　　　(b)

Figure 1. Bleached birch Kraft pulp fibers in an optical microscope (**a**) after ammonium persulfate (APS) treatment; (**b**) after APS treatment followed by three cycles of mechanical treatment in a high shear mixer and ultrasound treatment.

The reaction of cellulose and APS took place when persulfate ions $S_2O_8^{2-}$ were activated by the heating of the APS-cellulose solution to 70 °C in acidic conditions. Thus, $SO_4 \bullet^-$ was created in the mixture according to the previously reported reaction mechanism [27,48]. The proposed further reaction course involves the presence of other radicals and oxidants, including hydrogen peroxide. Both hydrogen peroxide and radicals penetrate the amorphous regions of cellulose and break down the disordered cellulose by co-hydrolyzing the 1,4-β bonds of the cellulose polymer chain [26], causing the observed loss of cellulose mass. Almost all authors who have worked with this method previously reported depolymerization of cellulose by APS treatment [26,29–31,33,35,38,39,41,43–46,49–51].

It was found that the depolymerization process of cellulose depends on the concentration of APS solution, treatment time [43] and mainly on the type of cellulose source [26]. The majority of published research on APS treatment of cellulose materials have focused on the total destruction of amorphous regions in order to obtain CNC. This research was devoted to the partial destruction of amorphous cellulose; therefore, a shorter reaction time was chosen in order to partly preserve the natural structure of cellulose. The partly depolymerized cellulose was used as a raw material for CNF production employing mechanical treatment.

The processing of cellulose was continued by seven cycles of the mechanical and ultrasound treatment. The ultrasound treatment was chosen as an intermediate step between the mechanical treatments in a high shear mixer to create additional hydrodynamic shear forces, which can contribute to the delamination process of cellulose fiber. MFC production by APS oxidation combined with ultrasonication was reported previously by Goh et al. [44]. The visualization of material after APS oxidation and mechanical treatments (CNF obtained by APS and mechanical treatment, hereinafter APS CNF) demonstrated delamination of cellulose fiber (Figure 1b). It was observed that after three cycles of mechanical treatment, individual micro- and nanofibrils can be seen separating from the bulk fiber. The level of fibrillation increased when the number of the treatment cycles was increased, leading to the high level of fibrillation seen after seven cycles of the treatment. The occurrence of CNF was identified by the increase in the apparent viscosity of the cellulose mixture and by AFM imaging. It was found that the obtained material consisted of a micro- and nanofibril mixture (Figure 2a), and objects with diameters in range from 20 nm to 300 nm were identified (Figure 2b). It is known that the presence of CNF can be proved by observing cellulose with lateral dimensions in the nanometer range [10]. This is also in accordance with the typical diameters of CNF reported by other authors [44]. The length of the fibrils was not detectable because of the entanglements. The presence of micro and macro objects in CNF correlates to the statement of Desmaisons et al., who reported that despite the fact that a CNF suspension in reality is composed of nano, micro and macro dimensions caused by incomplete defibrillation during mechanical shearing, it still can meet the quality standards of CNF [52].

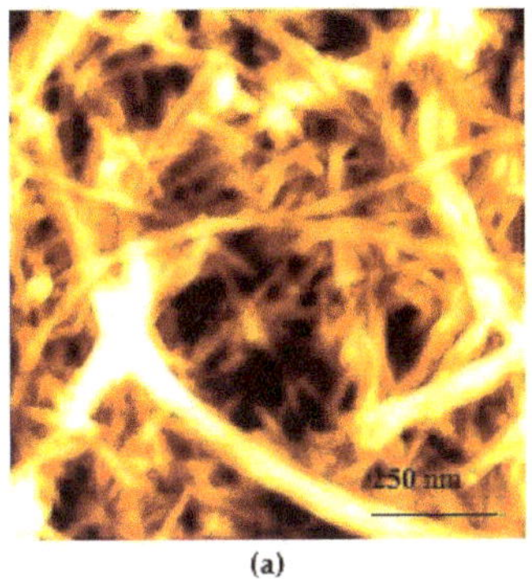

(a)

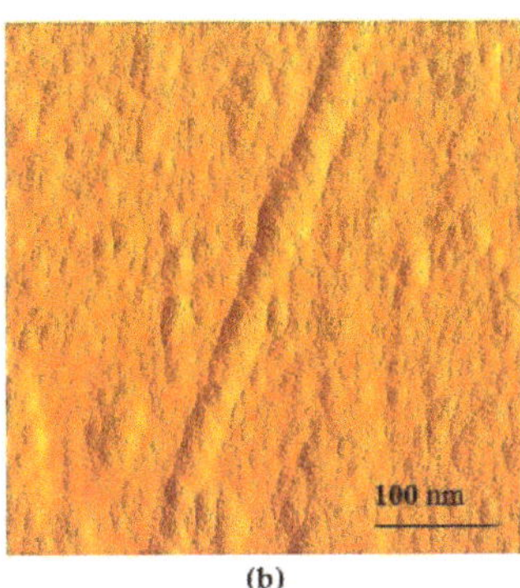

(b)

Figure 2. Atomic force microscopy (AFM) pictures of cellulose nanofibers (CNF) obtained from bleached birch Kraft pulp by combined chemical, mechanical and ultrasound treatment (a) a bundle of CNF; (b) a single fiber of CNF.

It is known that obtaining CNF is a high energy demanding process because of the use of high-pressure homogenizers, microfluidizers, grinders etc. [10]. The high energy demand has been a challenge for decades and the main barrier for commercial success until the strong impact of pretreatment methods was discovered, such as TEMPO oxidation or enzymatic hydrolysis. It has been identified that more focus should be devoted to the development of alternative pretreatment methods, which can benefit the production process or endow the CNF with new properties [10]. The goal of this research agrees with the previous statement because APS oxidation is used as a pretreatment method prior to the mechanical treatment of cellulose. A similar approach, with an oxidation step before rather mild mechanical treatment, has been already used before. It was reported that it is possible to obtain CNF by blending 0.7–2.0 wt % oxidized cellulose suspensions with conventional blenders [53,54].

3.2. FTIR Results

The FTIR spectra of neat cellulose and APS CNF are presented in Figure 3a. For both samples, the peak observed at ~3400 cm^{-1} is attributed to O–H vibration, mainly caused by hydrogen bonds in the cellulose. The peak at ~2900 cm^{-1} is associated with C–H stretching vibrations, and the peak at ~1640 cm^{-1} is related to the absorbed water. The peak at ~1375 cm^{-1} is associated with asymmetric vibrations of C–H band; and sharp peaks at 1060 cm^{-1} are associated with C–O stretching vibrations. Both cellulose and APS CNF exhibited similar peaks (Figure 3a), with exception of a peak at 1740 cm^{-1} for APS CNF (Figure 3b), which is related to the C=O stretching vibration. The presence of C=O confirmed that the hydroxyl groups of the CNF had been oxidized to carboxyl groups during

APS treatment. Results correlate with other authors [38,51] who reported that the hydroxyl groups at C6 of cellulose were regioselectively oxidized into carbonyl groups during APS treatment. The charges on the oxidized cellulose fibers enhance the swelling of the cellulose by electrostatic repulsion to facilitate the fibrillation into nanofibers [20]. Furthermore, it was reported that the presence of carbonyl groups on the surface of cellulose nanoparticles would make them more reactive and improve their properties needed for their use in nanocomposites [43], or make them more accessible for chemical derivatization, such as grafting [55].

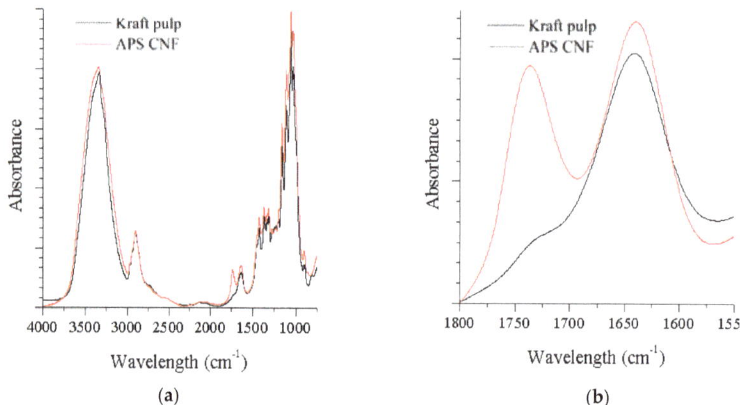

Figure 3. FTIR spectra of ammonium persulfate treated nanofibrillated cellulose compared to microcrystalline cellulose (**a**) spectra 4000–850 cm^{-1}; (**b**) spectra 1800–1550 cm^{-1}.

3.3. Thermal Stability

The thermal stability of the obtained APS CNF was investigated by TGA analysis. The TGA and differential thermogravimetry (DTG) curves of APS CNF in comparison to initial Kraft pulp are shown in Figure 4. The weight loss at temperatures below 120 °C corresponds to the evaporation of adsorbed water [33].

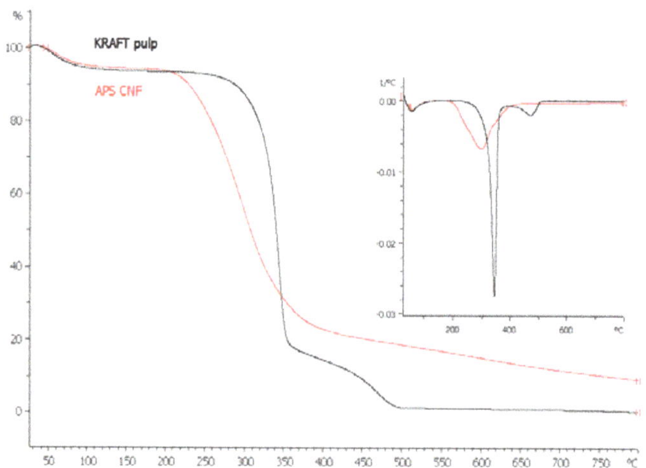

Figure 4. TG (big scale) and differential thermogravimetry DTG (small scale) curves of Kraft pulp and ammonium persulfate treated nanofibrillated cellulose (APS CNF).

Kraft pulp displayed the typical degradation profile of cellulose [56] with an onset temperature of 306 °C, while thermal degradation of APS CNF started at 225 °C, which was significantly lower than that of the initial pulp. Based on the DTG curves, the thermal decomposition peaks for the maximum weight loss of the Kraft pulp and APS CNF were at 347 °C and 298 °C, respectively. The thermal stability of cellulose material decreased after APS and mechanical treatments, most likely due to the mechanical degradation of cellulose [57]. The decreased degradation temperature of APS CNF could be related to the presence of carboxylic groups [46] and the structure of nanofibrillated cellulose, which have increased surface area compared to the Kraft pulp. Results were in good accordance to other authors [29,33], who reported about lower thermal stability of nanocellulose obtained by ammonium persulfate oxidation compared to the initial cellulosic material.

3.4. PXRD Crystallinity

The powder X-ray diffraction studies were conducted to estimate the crystallinity of obtained CNF. Figure 5 gives the PXRD spectra of APS CNF compared to the initial bleached Kraft pulp. As can be observed, both samples showed three characteristic peaks at 16°, 22.7° and 34.5°, representing the (110), (200) and (004) crystallographic planes of the monoclinic cellulose I lattice [58]. Crystallinity of the cellulose sample was calculated using Rietveld refinement on the obtained PXRD patterns in TOPAS v5. The crystallinity index (CI) of APS CNF was 74.3%, which was higher than the CI of the initial bleached Kraft pulp (69.6%). The higher CI of APS CNF indicated that part of the amorphous region in cellulose had been removed during APS treatment. A similar correlation was also found by Goh et al. [44], where oil palm empty fruit bunches were used as a raw material for obtaining MFC. The raw material had a CI of 74%, which was increased to 75% of MFC by APS pretreatment and ultrasonication. According to the investigations of Chen et al., the ultrasonic treatment had an insignificant impact on the crystal regions of the cellulose fibers [59]. We suggest that changes in the CI of APS CNF were made mainly by chemical pretreatment, while mechanical treatment left the CI of cellulose unchanged.

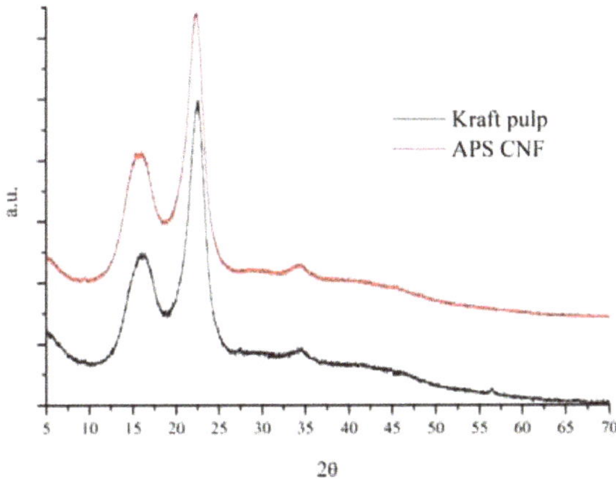

Figure 5. Powder X-ray diffraction spectra of persulfate treated nanofibrillated cellulose (APS CNF) and Kraft pulp.

The proposed method can be used to produce CNF with a defined CI, using pretreatment with APS solution to increase the CI of cellulose and a high shear mixer to defibrillate fibers. It has been reported by Jiang et al. that APS oxidation of cellulose material is easily controllable [38]. However, the majority of researchers have used APS treatment to disrupt the amorphous regions of the cellulose to reach the highest CI in order to obtain CNC. By employing APS oxidation with a duration of 12–16 h, it has been possible to produce CNC with a CI in range from 86.8% [51], 87% [38] and up to 93.9% [43].

3.5. Zeta Potential

The nanocellulose suspension is considered stable when the absolute value of the Zeta potential is higher than -25 mV [60,61]. The measurements of the Zeta potential of APS CNF were carried out after each cycle of the mechanical treatment. The average number was -26.9 ± 1.8 mV. It was identified that processing time and the number of cycles had no impact on the Zeta potential. It can be concluded that the required value of Zeta potential in order to obtain a stable suspension has been achieved in case of the APS CNF.

4. Conclusions

An innovative method for the production of cellulose nanofibrils with a diameter 20–300 nm by ammonium persulfate and further defibrillation using consequent mechanical treatment in a high shear laboratory mixer and ultrasonication was developed. It was found that the suspension of ammonium persulfate treated cellulose nanofibrils consists of different fractions, including nanofibrils, microfibrils, bundles and others. However, the suspension was considered stable because the Zeta potential value was -26.9 ± 1.8 mV. The crystallinity index of ammonium persulfate treated nanofibrillated cellulose was 74.3%, which was higher than the crystallinity index of the initial bleached Kraft pulp (69.6%). The thermal stability of cellulose material decreased after ammonium persulfate and mechanical treatments. The proposed method can be used to produce cellulose nanofibrils with defined crystallinity, using pretreatment with ammonium persulfate solution to increase the crystallinity index of cellulose and a high shear mixer and ultrasonication to defibrillate the fibers. The homogeneity of the produced cellulose nanofibril suspension is the main challenge for this method.

Author Contributions: Formal analysis and investigation, I.F., V.F.; investigation, A.B.; writing-original draft preparation, I.F.; writing-review and editing, U.C.

Funding: This work has been financed by the European Regional Development Fund Contract No. 1.1.1.1/16/A/031 "Rigid Polyurethane/Polyisocyanurate Foam Thermal Insulation Material Reinforced with Nano/MicroSize Cellulose".

Acknowledgments: Special acknowledgments to Andris Actins from the University of Latvia, Faculty of Chemistry, Riga, Latvia who performed XRD analysis and Andrejs Ogurcovs from Daugavpils University, Daugavpils, Latvia, who performed AFM imaging.

Conflicts of Interest: The authors declare no conflict of interest.

References

1. Ioelovich, M. Nanocellulose—Fabrication, structure, properties, and application in the area of care and cure. In *Fabrication and Self-Assembly of Nanobiomaterials, Applications of Nanobiomaterials*; Grumezescu, A.M., Ed.; William Andrew (Elsevier): Oxford, UK, 2016; Volume 1, pp. 243–288. ISBN 978-0-323-41533-0.
2. Mishra, R.K.; Sabu, A.; Tiwari, S.K. Materials chemistry and the futurist eco-friendly applications of nanocellulose: Status and prospect. *J. Saudi Chem. Soc.* **2018**, in press. Available online: https://doi.org/10.1016/j.jscs.2018.02.005 (accessed on 10 August 2018).
3. Kargarzadeh, H.; Mariano, M.; Huang, J.; Lin, N.; Ahmad, I.; Dufresne, A.; Thomas, S. Recent developments on nanocellulose reinforced polymer nanocomposites: A review. *Polymer* **2017**, *132*, 368–393. [CrossRef]
4. Mondal, S. Preparation, properties and applications of nanocellulosic materials. *Carbohydr. Polym.* **2017**, *163*, 301–316. [CrossRef] [PubMed]
5. Abitbol, T.; Rivkin, A.; Cao, Y.; Nevo, Y.; Abraham, E.; Ben-Shalom, T.; Lapidot, S.; Shoseyov, O. Nanocellulose, a tiny fiber with huge applications. *Curr. Opin. Biotechnol.* **2016**, *39*, 76–88. [CrossRef] [PubMed]
6. Dufresne, A. Nanocellulose: A new ageless bionanomaterial. *Mater. Today* **2013**, *16*, 220–227. [CrossRef]
7. Deepa, B.; Abraham, E.; Cordeiro, N.; Mozetic, M.; Mathew, A.P.; Oksman, K.; Faria, M.; Thomas, S.; Pothan, L.A. Utilization of various lignocellulosic biomass for the production of nanocellulose: A comparative study. *Cellulose* **2015**, *22*, 1075–1090. [CrossRef]

8. Klemm, D.; Kramer, F.; Moritz, S.; Lindstrom, T.; Ankerfors, M.; Gray, D.; Dorris, A. Nanocelluloses: A new family of nature based materials. *Angew. Chem. Int. Ed.* **2011**, *50*, 5438–5466. [CrossRef] [PubMed]

9. Peng, B.L.; Dhar, N.; Liu, H.L.; Tam, K.C. Chemistry and applications of nanocrystalline cellulose and its derivatives: A nanotechnology perspective. *Can. J. Chem. Eng.* **2011**, *89*, 1191–1206. [CrossRef]

10. Nechyporchuk, O.; Belgacem, M.N.; Bras, J. Production of cellulose nanofibrils: A review of recent advances. *Ind. Crop. Prod.* **2016**, *93*, 2–25. [CrossRef]

11. Zhang, Y.; Nypelö, T.; Salas, C.; Arboleda, J.; Hoeger, I.C.; Rojas, O.J. Cellulose nanofibrils: From strong materials to bioactive surfaces. *J. Renew. Mater.* **2013**, *1*, 195–211. [CrossRef]

12. Khalil, H.P.S.A.; Bhat, A.H.; Yusra, A.F.I. Green composites from sustainable cellulose nanofibrils: A review. *Carbohydr. Polym.* **2012**, *87*, 963–979. [CrossRef]

13. Lavoine, N.; Bergström, L. Nanocellulose-based foams and aerogels: Processing, properties, and applications. *J. Mater. Chem. A* **2017**, *5*, 16105–16117. [CrossRef]

14. Ferrer, A.; Pal, L.; Hubbe, M. Nanocellulose in packaging: Advances in barrier layer technologies. *Ind. Crop. Prod.* **2017**, *95*, 574–582. [CrossRef]

15. Ramírez, J.A.Á.; Fortunati, E.; Kenny, J.M.; Torre, L.; Foresti, M.L. Simple citric acid-catalyzed surface esterification of cellulose nanocrystals. *Carbohydr. Polym.* **2017**, *157*, 1358–1364. [CrossRef] [PubMed]

16. Ruiz-Palomero, C.; Soriano, M.L.; Valcárcel, M. Nanocellulose as analyte and analytical tool: Opportunities and challenges. *TrAC Trend Anal. Chem.* **2017**, *87*, 1–18. [CrossRef]

17. Filpponen, I.; Kontturi, E.; Nummelin, S.; Rosilo, H.; Kolehmainen, E.; Ikkala, O.; Laine, J. Generic method for modular surface modification of cellulosic materials in aqueous medium by sequential "Click" reaction and adsorption. *Biomacromolecules* **2012**, *13*, 736–742. [CrossRef] [PubMed]

18. Tang, J.; Sisler, J.; Grishkewich, N.; Tam, K.C. Functionalization of cellulose nanocrystals for advanced applications. *J. Colloid Interface Sci.* **2017**, *494*, 397–409. [CrossRef] [PubMed]

19. Coseri, S. Cellulose: To depolymerize . . . or not to? *Biotechnol. Adv.* **2017**, *35*, 251–266. [CrossRef] [PubMed]

20. Isogai, A.; Saito, T.; Fukuzumi, H. TEMPO-oxidizes cellulose nanofibers. *Nanoscale* **2011**, *3*, 71–85. [CrossRef] [PubMed]

21. Asad, M.; Saba, N.; Asiri, A.M.; Jawaid, M.; Wanrosli, W.D. Preparation and characterization of nanocomposite films from oil palm pulp nanocellulose/poly (Vinyl alcohol) by casting method. *Carbohydr. Polym.* **2018**, *191*, 103–111. [CrossRef] [PubMed]

22. Zhang, S.; Feng, J.; Feng, J.; Jiang, Y.; Ding, F. Carbon aerogels by pyrolysis of TEMPO-oxidized cellulose. *Appl. Surf. Sci.* **2018**, *440*, 873–879. [CrossRef]

23. Zhang, K.; Shen, M.; Liu, H.; Shang, S.; Wang, D.; Liimatainen, H. Facile synthesis of palladium and gold nanoparticles by using dialdehyde nanocellulose as template and reducing agent. *Carbohydr. Polym.* **2018**, *186*, 132–139. [CrossRef] [PubMed]

24. Hamou, K.B.; Kaddami, H.; Dufresne, A.; Boufi, S.; Magnin, A.; Erchiqui, F. Impact of TEMPO-oxidation strength on the properties of cellulose nanofibril reinforced polyvinyl acetate nanocomposites. *Carbohydr. Polym.* **2018**, *181*, 1061–1070. [CrossRef] [PubMed]

25. Mhd Haniffa, M.A.C.; Ching, Y.C.; Chuah, C.H.; Ching, K.Y.; Nai-Shang, L. Effect of TEMPO-oxidization and rapid cooling on thermo-structural properties of nanocellulose. *Carbohydr. Polym.* **2017**, *173*, 91–99. [CrossRef] [PubMed]

26. Leung, A.C.W.; Hrapovic, S.; Lam, E.; Liu, Y.; Male, H.B.; Mahmoud, K.A.; Luong, J.H.T. Characteristics and properties of carboxylated cellulose nanocrystals prepared form a novel one-step procedure. *Small* **2011**, *7*, 302–305. [CrossRef] [PubMed]

27. Waldemer, R.H.; Tratnyek, P.G.; Johnson, R.L.; Nurmi, A.J. Oxidation of Chlorinated Ethenes by Heat-Activated Persulfate: Kinetics and Products. *Environ. Sci. Technol.* **2007**, *41*, 1010–1015. [CrossRef] [PubMed]

28. Trache, D.; Hussin, M.H.; Haafizc, M.K.M.; Thakur, V.K. Recent progress in cellulose nanocrystals: Sources and production. *Nanoscale* **2017**, *9*, 1763–1786. [CrossRef] [PubMed]

29. Zhang, K.; Sun, P.; Liu, H.; Shang, S.; Song, J.; Wang, D. Extraction and comparison of carboxylated cellulose nanocrystals from bleached sugarcane bagasse pulp using two different oxidation methods. *Carbohydr. Polym.* **2016**, *138*, 237–243. [CrossRef] [PubMed]

30. Castro-Guerrero, C.F.; Gray, D.G. Chiral nematic phase formation by aqueous suspensions of cellulose nanocrystals prepared by oxidation with ammonium persulfate. *Cellulose* **2014**, *21*, 2567–2577. [CrossRef]

31. Mascheroni, E.; Rampazzo, R.; Ortenzi, M.A.; Piva, G.; Bonetti, S.; Piergiovanni, L. Comparison of cellulose nanocrystals obtained by sulfuric acid hydrolysis and ammonium persulfate, to be used as coating on flexible food-packaging materials. *Cellulose* **2016**, *23*, 779–793. [CrossRef]

32. Rampazzo, R.; Alkan, D.; Gazzotti, S.; Ortenzi, M.A.; Piva, G.; Piergiovanni, L. Cellulose Nanocrystals form lignocelllosics Raw materials, for oxygen barrier coatings on food packaging films. *Packag. Technol. Sci.* **2017**, *30*, 645–661. [CrossRef]

33. Oun, A.A.; Rhim, J.W. Characterization of carboxymethyl cellulose-based nanocomposite films reinforced with oxidized nanocellulose isolated using ammonium persulfate method. *Carbohydr. Polym.* **2017**, *174*, 484–492. [CrossRef] [PubMed]

34. Male, K.B.; Leung, A.C.W.; Montes, J.; Kamen, A.; Luong, J.H.T. Probing inhibitory effects of nanocrystalline cellulose: Inhibition versus surface charge. *Nanoscale* **2012**, *4*, 1373–1379. [CrossRef] [PubMed]

35. Lam, E.; Male, K.B.; Chong, J.H.; Leung, A.C.W.; Luong, J.H.T. Applications of functionalized and nanoparticle-modified nanocrystalline cellulose. *Trends Biotechnol.* **2012**, *30*, 283–290. [CrossRef] [PubMed]

36. Xu, G.; Liang, S.; Zhang, M.; Fan, J.; Yu, X. Studies on the electrochemical and dopamine sensing properties of AgNP-modified carboxylated cellulose nanocrystal-doped poly (3,4-ethylenedioxythiophene). *Ionics* **2017**, *23*, 3211–3218. [CrossRef]

37. Vecbiskena, L.; Rozenberga, L. Nanocelluloses obtained by ammonium persulfate (APS) oxidation of bleached kraft pulp (BKP) and bacterial cellulose (BC) and their application in biocomposite films together with chitosan. *Holzforschung* **2017**, *71*, 659–666. [CrossRef]

38. Jiang, H.; Wu, Y.; Han, B.; Zhang, Y. Effect of oxidation time on the properties of cellulose nanocrystals from hybrid poplar residues using the ammonium persulfate. *Carbohydr. Polym.* **2017**, *174*, 291–298. [CrossRef] [PubMed]

39. Rozenberga, L.; Skute, M.; Belkova, L.; Sable, I.; Vikele, L.; Semjonovs, P.; Saka, M.; Ruklisha, M.; Paegle, L. Characterisation of films and nanopaper obtained from cellulose synthesised by acetic acid bacteria. *Carbohydr. Polym.* **2016**, *144*, 33–40. [CrossRef] [PubMed]

40. Pirich, C.L.; de Freitas, R.A.; Torresi, R.M.; Picheth, G.F.; Sierakowski, M.R. Piezoelectric immunochip coated with thin films of bacterial cellulose nanocrystals for dengue detection. *Biosens. Bioelectron.* **2017**, *92*, 47–53. [CrossRef] [PubMed]

41. Hu, Y.; Tang, L.; Lu, Q.; Wang, S.; Chen, X.; Huang, B. Preparation of cellulose nanocrystals and carboxylated cellulose nanocrystals from borer powder of bamboo. *Cellulose* **2014**, *21*, 1611–1618. [CrossRef]

42. Wen, C.; Yuan, Q.; Liang, H.; Vriesekoop, F. Preparation and stabilization of D-limonene Pickering emulsions by cellulose nanocrystals. *Carbohydr. Polym.* **2014**, *112*, 695–700. [CrossRef] [PubMed]

43. Cheng, M.; Qin, Z.; Liu, Y.; Qin, Y.; Li, T.; Chen, L.; Zhua, M. Efficient extraction of carboxylated spherical cellulose nanocrystals with narrow distribution through hydrolysis of lyocell fibers by using ammonium persulfate as an oxidant. *J. Mater. Chem. A* **2014**, *2*, 251–258. [CrossRef]

44. Goh, K.Y.; Ching, Y.C.; Chuah, C.H.; Abdullah, C.L.; Liou, N.-S. Individualization of microfibrillated celluloses from oil palm empty fruit bunch: Comparative studies between acid hydrolysis and ammonium persulfate oxidation. *Cellulose* **2016**, *23*, 379–390. [CrossRef]

45. Nascimento, D.M.D.; Almeida, J.S.; do Vale, S.M.; Leitão, R.C.; Muniz, C.R.; de Figueirêdo, M.C.B.; Morais, J.P.S.; Rosa, M.D.F. A comprehensive approach for obtaining cellulose nanocrystal from coconut fiber. Part I: Proposition of technological pathways. *Ind. Crop. Prod.* **2016**, *93*, 66–75. [CrossRef]

46. Du, C.; Liu, M.; Li, B.; Li, H.; Meng, Q.; Zhan, H. Cellulose Nanocrystals Prepared by persulfate one-step oxidation of bleached bagasse pulp. *Bioresources* **2016**, *11*, 4017–4024. [CrossRef]

47. Scarlett, N.V.Y.; Madsen, I.C. Quantification of phases with partial or no known crystal structures. *Powder Diffr.* **2006**, *21*, 278–284. [CrossRef]

48. Manz, K.E.; Carter, K.E. Investigating the effects of heat activated persulfate on the degradation of furfural, a component of hydraulic fracturing fluid chemical additives. *Chem. Eng. J.* **2017**, *327*, 1021–1032. [CrossRef]

49. Lam, E.; Leung, A.C.W.; Liu, Y.; Majid, E.; Hrapovic, S.; Male, K.B.; Luong, J.H.T. Green Strategy Guided by Raman Spectroscopy for the Synthesis of Ammonium Carboxylated Nanocrystalline Cellulose and the Recovery of Byproducts. *ACS Sustain. Chem. Eng.* **2013**, *1*, 278–283. [CrossRef]

50. Rozenberga, L.; Vikele, L.; Vecbiskena, L.; Sable, I.; Laka, M.; Grinfelds, U. Preparation of nanocellulose using ammonium persulfate and method's comparison with other techniques. *Key Eng. Mater.* **2016**, *674*, 21–25. [CrossRef]

51. Wu, Y.; Cao, F.; Jiang, H.; Zhang, Y. Carboxyl cellulose nanocrystal extracted from hybrid poplar residue. *Bioresorces* **2017**, *12*, 8775–8785. [CrossRef]

52. Desmaisons, J.; Boutonnet, E.; Rueff, M.; Dufresne, A.; Bras, J. A new quality index for benchmarking of different cellulose nanofibrils. *Carbohydr. Polym.* **2017**, *174*, 318–329. [CrossRef] [PubMed]

53. Uetani, K.; Yano, H. Nanofibrillation of Wood Pulp Using a High-Speed Blender. *Biomacromolecules* **2011**, *12*, 348–353. [CrossRef] [PubMed]

54. Boufi, S.; Chaker, A. Easy production of cellulose nanofibrils from corn stalk by a conventional high speed blender. *Ind. Crop. Prod.* **2016**, *93*, 39–47. [CrossRef]

55. Benkaddour, A.; Jradi, K.; Robert, S.; Daneult, C. Grafting of Polycaprolactone on Oxidized Nanocelluloses by Click Chemistry. *Nanomaterials* **2013**, *3*, 141–157. [CrossRef] [PubMed]

56. Yang, X.; Zhao, Y.; Li, R.; Wu, Y.; Yang, M.A. modified kinetic analysis method of cellulose pyrolysis based on TG–FTIR technique. *Thermochim. Acta* **2018**, *665*, 20–27. [CrossRef]

57. Wang, W.; Liang, T.; Bai, H.; Dong, W.; Liu, X. All cellulose composites based on cellulose diacetate and nanofibrillated cellulose prepared by alkali treatment. *Carbohydr. Polym.* **2018**, *179*, 297–304. [CrossRef] [PubMed]

58. Xing, L.; Gu, J.; Zhang, W.; Tu, D.; Hu, C. Cellulose I and II nanocrystals produced by sulfuric acid hydrolysis of Tetra pak cellulose I. *Carbohydr. Polym.* **2018**, *192*, 184–192. [CrossRef] [PubMed]

59. Chen, W.; Haipeng, W.; Liu, Y.; Chen, P.; Zhang, M.; Hi, Y. Individualization of cellulose nanofibers from wood using high-intensity ultrasonication combined with chemical pretreatments. *Carbohydr. Polym.* **2011**, *83*, 1804–1811. [CrossRef]

60. Morais, J.P.S.; Rosa, M.F.; Filho, M.M.S.; Nascimento, L.D.; Nascimento, D.M.; Cassales, A.R. Extraction and characterization of nanocellulose structures from raw cotton linter. *Carbohydr. Polym.* **2013**, *91*, 229–235. [CrossRef] [PubMed]

61. Mirhosseini, H.; Tan, C.P.; Hamid, N.S.A.; Yusof, S. Effect of Arabic gum, xanthan gum and orange oil contents on zeta-potential, conductivity, stability, size index and pH of orange beverage emulsion. *Colloid Surf. A* **2008**, *315*, 47–56. [CrossRef]

 *nanomaterials*

Review

Bacterial Cellulose: Production, Modification and Perspectives in Biomedical Applications

Selestina Gorgieva [1,2,*] and **Janja Trček** [3,4]

1   Faculty of Mechanical Engineering, Institute of Engineering Materials and Design, University of Maribor, 2000 Maribor, Slovenia
2   Faculty of Electrical Engineering and Computer Science, Institute of Automation, University of Maribor, 2000 Maribor, Slovenia
3   Faculty of Natural Sciences and Mathematics, Department of Biology, University of Maribor, 2000 Maribor, Slovenia; janja.trcek@um.si
4   Faculty of Chemistry and Chemical Engineering, University of Maribor, 2000 Maribor, Slovenia
*   Correspondence: selestina.gorgieva@um.si; Tel.: +386-2220-7740; Fax: +386-2220-7990

Received: 26 August 2019; Accepted: 16 September 2019; Published: 20 September 2019

Abstract: Bacterial cellulose (BC) is ultrafine, nanofibrillar material with an exclusive combination of properties such as high crystallinity (84%–89%) and polymerization degree, high surface area (high aspect ratio of fibers with diameter 20–100 nm), high flexibility and tensile strength (Young modulus of 15–18 GPa), high water-holding capacity (over 100 times of its own weight), etc. Due to high purity, i.e., absence of lignin and hemicellulose, BC is considered as a non-cytotoxic, non-genotoxic and highly biocompatible material, attracting interest in diverse areas with hallmarks in medicine. The presented review summarizes the microbial aspects of BC production (bacterial strains, carbon sources and media) and versatile in situ and ex situ methods applied in BC modification, especially towards bionic design for applications in regenerative medicine, from wound healing and artificial skin, blood vessels, coverings in nerve surgery, dura mater prosthesis, arterial stent coating, cartilage and bone repair implants, etc. The paper concludes with challenges and perspectives in light of further translation in highly valuable medical products.

Keywords: bacterial cellulose; carbon source; in situ modification; ex situ modification; biomedical applications

---

1. Introduction

Cellulose is one of the most abundant biopolymers on Earth, and is mainly of plant, wood and bacterial origin. The cellulose of bacterial origin exhibits the highest purity and has thus attracted the interest of many researchers and industrial sectors. Generally, it consists of randomly assembled, <100 nm wide ribbon-shaped fibrils, composed of 7–8 nm-wide elementary nanofibrils aggregated in bundles. As such, it delivers a combination of exclusive properties, such as flexibility, high water holding capacity, hydrophilicity, crystallinity, mouldability in different shapes, elevated purity with absence of lignin and hemicellulose and biomimetic three-dimensional (3D) network as a hallmark. Because of these features, this type of cellulose attracts interest for different medical applications such as the engineering of artificial skin (particularly in recuperation of burned skin), artificial blood vessels, topical covering for severe wounds, coverings in nerve surgery, dura mater prosthesis, arterial stent coating, wound dressings, hemostatic material, electronic platforms, implants for cartilage and bone repair etc.

For efficient bacterial cellulose (BC) production we need an efficient and stabile bacterial strain with demands for growing that are not too expensive and with ability of being easily scaled up to industrial settings. The produced cellulose is generally easily separated from growth medium and further on modified using different approaches for various medically relevant applications. All these aspects (Figure 1) will be discussed in this review paper.

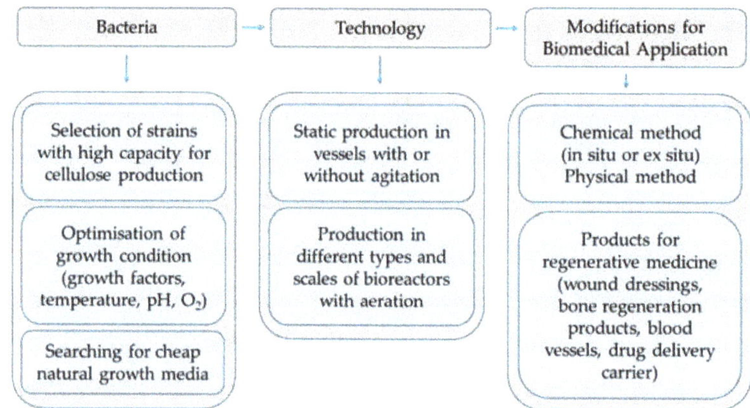

Figure 1. Scheme presenting the most important aspects which have to be considered for bacterial cellulose (BC)-production with biomedical application.

2. Bacteria Have High Capacity for Cellulose Production

BC is nanofibrillar, extracellular polysaccharide produced by diverse bacteria when they are growing statically, but also when bacteria are submerged in liquid and cultured by shaking. Bacteria produce BC in media with different carbon sources, although the efficiency of BC production differs substantially among various growth substrates. The substrate supplies energy to bacterial metabolism during the exhaustive energy-consuming pathway of cellulose synthesis. Theoretically, every carbon block which the bacterial cell metabolizes into glucose, can be used for cellulose production [1,2].

The capacity of BC production is widespread among bacteria, but the most prominent and well-known BC-producer is species *Komagataeibacter xylinus*, which belongs to the group of acetic acid bacteria (AAB). AAB are strictly aerobic Gram-negative bacteria classified into α-*Proteobacteria* [3–5]. The species has been for many years known as *Acetobacter xylinum*, but has been later classified into *Gluconacetobacter xylinus* and due to further taxonomic changes finally reclassified into *Komagataeibacter xylinus*. *K. xylinus* is not the only species among AAB with an immense potential for BC production, since also other species, such as *Komagataeibacter hansenii*, *Komagataeibacter medellinensis*, *Komagataeibacter nataicola*, *Komagataeibacter oboediens*, *Komagataeibacter rhaeticus*, *Komagataeibacter saccharivorans* and *Komagataeibacter pomaceti* have been characterized as strong cellulose producers [4,6–9]. An important aspect of using AAB for cellulose production is their characteristic of being food-grade or GRAS bacteria (generally recognized as safe).

BC is synthesized in bacterial membrane from nucleotide-activated glucose [10]. Bacteria then channel BC through pores of cell membrane as fibrils composed of D-glucose units which are linked with β-1,4-glycosidic bonds. The chain is linear and extruded from the cell. Then the lateral and unidirectional aligned chains form intra- and inter-chain hydrogen bonding through all available hydroxyl groups. In this way the chains merge into insoluble nanofibrils of up to 25 nm in width and 1 to 9 μm in length which represents 2000 to 18,000 glucose residues [11]. These nanofibrils further aggregate into <100 nm wide ribbon-shaped fibrils what delivers a combination of exclusive properties to BC such as high water holding capacity, hydrophilicity, crystallinity and mouldability. Although almost all hydroxyl groups of the cellulose polymer are occupied with hydrogen bonds, one end of each cellulose polymer carries an unmodified C4-hydroxyl group and the opposite end a free C1-hydroxyl group, both of them representing possible sites for chemical modifications of cellulose [12].

Synthesis of nucleotide-activated glucose takes place in bacterial cytoplasm. If the starting substrate is glucose, the uridine diphosphate (UDP)-glucose is produced in three steps: phosphorylation of glucose by glucokinase, isomerization of glucose-6-phosphate into glucose-1-phosphate by phosphoglucomutase and synthesis of UDP-glucose by uridylyltransferase (UTP)-glucose-1-phosphate. Finally, cellulose synthase transfers glucosyl residues from UDP-glucose to the nascent β-D-1,4-glucan chain. Cellulose synthase is a membrane-embedded glycosyltransferase composed of two or three subunits [13]. The catalytic subunit of cellulose synthase is a major determinant of chemical and physical properties of BC, meaning that different bacterial species are able to generate cellulose with different lengths [14].

Comparison of AAB genomes revealed that AAB can possess more operons for cellulose production, moreover, the composition of operons differs from each other [4,15]. These differences very likely influence cellulose synthesis, cellulose transport to the cell surface and/or assembly of fibrils into ribbons [1].

The BC is not the only extracellular polysaccharide secreted by AAB. The other two well-known extracellular polysaccharides, acetan and levan are, however, water soluble [4,16,17]. Interestingly, acetan was first described in species *Komagataeibacter xylinus*. In contract to cellulose, acetan is branched acidic heteropolysaccharide [18]. Ishida et al. [19] identified lower cellulose production in a mutant not producing acetan. However, the cellulose production could be recovered by addition of acetan into the medium, meaning that the synthesis of both polymers is not connected at the genetic level.

3. Different Carbon Sources Used for Bacterial Cellulose (BC) Production

The production of BC is extremely expensive, which is mainly a consequence of high costs of synthetic media for its production. The most well-known complex synthetic medium for growing cellulose producing AAB is Hestrin–Schramm medium (HS), composed of 2% (w/v) glucose, 0.5% (w/v) peptone, 0.5% (w/v) yeast extract, 0.27% (w/v) Na_2HPO_4 and 1.15 g/L citric acid [20]. During BC production, other by-products, such as gluconic and other acids are formed, that can decrease the BC yield [8]. The composition of HS medium can be further optimized for the highest cellulose yield by replacing glucose with other carbon sources, such as maltose, fructose, cellobiose, mannitol, xylose, sucrose, galactose etc. In most cases glucose turned out to be the best energy source for bacteria, besides, glucose can be directly used as precursor for the assembly of glucose units into cellulose. Wang et al. [2] have recently reported that fructose had in their microbial process higher cellulose yield in comparison to other carbon sources, also to glucose. The process for BC production can be further optimized by adding buffers into medium for keeping pH at optimal value for growing bacterial strains [6].

To reduce the costs for BC production, the alternative natural carbon sources are utilized, such as waste substrate from different sectors of the food industry, sugar cane molasses etc. The BC yield can be improved also by addition of additives into growth medium such as glycerol, agar, xanthan, sodium alginate, ethanol (Figure 2), carboxymethyl cellulose (CMC), etc. Naritomi et al. [21] reported on enhanced cellulose yield during continuous BC production with *K. xylinus* subsp. *sucrofermentans* BPR3OO1A using fructose medium supplemented with 0.1 wt% of ethanol. The production of cellulose in a static culture with strain *K. xylinus* DA increased about 4-fold as a result of adding 2 wt% acetic acid in glucose medium [22]. Lu et al. [23] reported enhanced BC production with *K. xylinus* in chemically defined medium under static cultivation by the addition of pyruvic acid, malic acid, lactic acid, acetic acid, citric acid, succinic acid, and ethanol (Figure 2) in concentrations 0.15%, 0.1%, 0.3%, 0.4%, 0.1%, 0.2%, 4%, respectively. Li et al. [24] improved cellulose production with the strain *K. hansenii* M2010332 by the addition of ethanol and sodium citrate. Lu et al. [25] reported that the addition of 1% of methanol, 0.5% ethylene glycol, 0.5% of n-propanol, 3% of glycerol, 0.5% of n-butanol and 4% of mannitol produced 21.8%, 24.1%, 13.4%, 27.4%, 56% and 47.3% higher yield of cellulose by culturing strain *K. xylinus* 186 statically in glucose medium. The experiments of Matsuoka et al. [26] showed that the addition of lactate and methionine in fructose medium improved cellulose production with

K. xylinus subsp. sucrofermentas BPR200. However, the BC yield reached 90% of that obtained in corn steep liquor. There is also a report on improved BC production with *K. xylinus* ATCC 10,245 by adding vitamin C in growth medium [27].

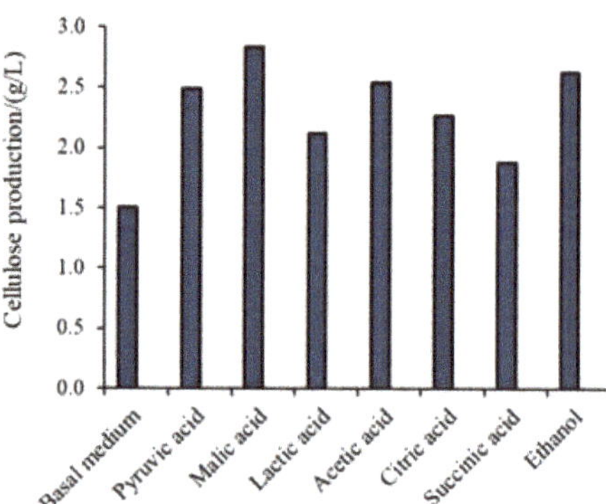

Figure 2. Influence of different organic acids and ethanol on cellulose yield. Reproduced from [21] with permission from Research & Reviews: Journal of Microbiology and Biotechnology, 2016.

The production of BC in synthetic media with different carbon sources and growth factors, which are usually added as yeast extract and peptone, is expensive. The researchers are thus searching for inexpensive raw material containing high levels of sugars as substrates for BC production. To this aim several raw materials have been analyzed for BC production, such as tobacco waste extract [22], sugar beet molasses, cheese whey media [23], distillery effluent [24], corn steep liquor [25], fruit juice [26], corn stalks [28], litchi extract [29], beverage industrial waste [30], corncob acid hydrolysate [31] and waste beer yeast [32]. Another possible natural growth medium would be waste material from wine production. According to recent reports [33], 1.17 kg of grapes are used to produce 750 mL wine, and after the grapes are squeezed, about 20% of that weight remains in the form of grape skins, seeds and stems, counting for ~12 million tons each year. This substrate contains soluble carbohydrates (white grapes), fibers, acids, salts, and phenolic compounds (red grapes) [34] and as such it is often considered as a convenient source of carbon for microbial processes. Moreover, grape waste as carbon source in BC production may contribute to reduce winery residuals, reduce BC production costs, offering new ways to diversify BC production by taking into account also the environmental aspect by diminishing waste products in nature.

The carbon source used for growing BC-producers affects BC properties: water holding capacity, surface area, porosity, polymerization degree, molecular weight, crystallinity index (67%–96%), mean crystallite size (5.7–6.4 nm), intrinsic viscosity, oxygen and water vapor transmission rates, mechanical properties, etc. Molina-Ramírez et al. [35] reported improved BC yield by addition of ethanol and acetic acid in growth medium, however, the crystallinity index, the degree of polymerization and maximum rate of degradation temperatures decreased by 9.2%, 36%, and 4.96%, respectively, by the addition of ethanol and by 7.2%, 27%, and 4.21%, respectively, by the addition of acetic acid. The crystallinity index of BC produced in the presence of ascorbic acid also decreased with remarkable change in d-spacing [27]. However, a recent study of Wang et al. [2] reported similar morphology and microfibrils of BCs from different carbon sources, meaning that these characteristics have to be checked for each bacterial strain before starting BC production at large scale.

The production of BC can be simply performed in vessels with large surface area which support direct supply of oxygen and assembly of large cellulose sheets (Figure 3). To improve the efficiency of BC production and to produce cellulose of desired characteristics, different technological approaches can be used (Table 1).

Figure 3. Static production of cellulose by *Komagataeibacter maltaceti* 1529[T] on complex microbiological medium.

Table 1. The most common methods for bacterial cellulose (BC) production.

Method for BC Production	Basic Characteristics of The Process and The Cellulose
Static production [36]	Most commonly used method at the lab scale. Duration of the process is up to two weeks. Cellulose is in the form of hydrogel sheet.
Production in shaking culture [37,38]	Increased delivery of oxygen to bacteria. Might result in reduced genetic stability of bacteria and lower BC production. Production of cellulose of different particle sizes and various shapes (mainly of spherical structure). Suitable for economic scale production.
Production in airlift bioreactor [38,39]	Efficient oxygen supply with low power supply. Cellulose produced in pellet.
Production in rotating disc bioreactors [40]	Production of homogenous cellulose. Cellulose yield is compared to the static process.
Production in trickling bed reactor [41]	Provides high oxygen concentration and low shear force. Produce BC in form of irregular sheets.

4. BC Modifications with Medical Relevance

3D structuring of BC within a translucent, gelatinous, interwoven, nano-fibrous network of linear polysaccharide polymers is formed at static conditions, as displayed within Figure 4. In comparison with vegetal cellulose sources, BC demonstrate remarkable mechanical properties, such as flexibility [42] and soft-tissue resembling stress-strain behavior [43], as well as a high level of crystallinity and water-holding capacity. BC is a very pure material where common cellulose associates, i.e., lignin and hemicellulose, are absent. As such, is considered a non-cytotoxic, non-genotoxic and highly biocompatible material.

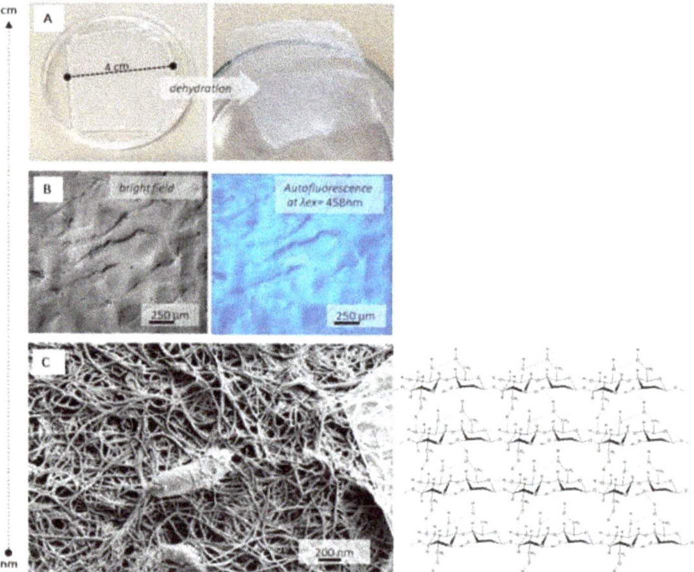

Figure 4. Different length-scale presentation of BC: (**A**) photographs of wet (left) and dry (right) BC membrane, (**B**) confocal fluorescent microscopy (CFM) image obtained under argon laser excitation at 458 nm from bright field and fluorescence channel, utilizing the cellulose autofluorescence and (**C**) high magnification scanning electron microscopy (SEM) image presenting entrapped *K. xylinus* bacteria and cellulose backbone insert.

However, BC lacks appropriate functionalities to trigger the initial cell attachment and control over the porosity, and it has very slow degradation, etc. To overcome this, BC has been modified by chemical (modification of chemical structure and functionalities) and physical means (change in porosity, crystallinity and fiber density) by applying versatile in situ and ex situ methods. In situ modifications are performed by the variation of culture media, carbon source and addition of other materials, while ex situ modifications are carried out by chemical and physical treatment of formed BC.

Chemical modification rely on inherent chemical reactivity due to the presence of hydroxyl groups, allowing reaction not only at heterogeneous, but also under homogeneous conditions. When compared with plant cellulose, the BC was found to be more reactive towards cynoethylation and carboxymethylation [44]. The homogeneous reaction including dissolving of BC with acetic anhydride and further iodination also reveals the highest reactivity of BC, yet, such a type of modification destroys the nanofibrillar structure.

Variation of water content within BC largely influences its viscoelastic and electrochemical properties. Due to increased resistance of BC to electron transfer, it becomes stiff at 50%–80% of water [45]. Such a finding was particularly important in wound dressing applications, where moisture content is an imperative. Addition of water-soluble polymers, such as CMC, methylcellulose (MC), and poly(vinyl alcohol) (PVA), was found to influence the water content of never dried and re-swollen BC [46]. On the other hand, Bottan et al. [47] introduced the guided assembly-based biolitography as technique to change the BC surface topography what is related to migratory patterns and alignments of human dermal cells, the fibroblasts and keratinocytes.

Some of modifications and resulting properties of BC are summarized within Table 2.

Table 2. Modifications of BC and resulting properties.

Modification	Application	Resulting Properties
BC nanocrystals/Regenerated Chitin fibers (BCNC/RC) [48]	Suture biomaterials	Biocompatible surgical sutures increasing strength of BCNC/RC filaments; Enzymatic degradation possible; Degradation rate can be tuned by varying concentration of BCNCs in the yarn; Chitin can promote cell proliferation (in vivo).
BC with modified topography [47]		Improved cell alignment; Promotion of fibroblast infiltration and new collagen deposition in the wound bed.
Vaccarin impregnated on BC [49]		Neovascularization; Stratified squamous epithelium; Dense new-born subcutaneous tissue formation of collagen fibers and hyperplastic fibrous connective tissue.
2,2,6,6-Tetramethylpiperidinyloxy (TEMPO)-Oxidized BC with Ag nanoparticles [50]		Antimicrobial activity; Ag^+ release with a rate of 12.2%/day at 37 °C in 3 days; Biocompatible.
BC/ZnO nanocomposite [51]	Wound dressing	Antimicrobial activity against *Escherichia coli*, *Pseudomonas aeruginosa*, *Staphylococcus aureus* and *Citrobacter freundii*; Significant healing of 66% after 15 days related to day 0.
BC/TiO$_2$ nanocomposite [52]		Antimicrobial activity against *Escherichia coli* and *Staphylococcus aureus*.
BC/ε -poly-L-Lysine (ε-PLL) nanocomposite [53]		Antimicrobial activity (broad-spectrum) without affecting the beneficial structural and mechanical properties; Modification with non-toxic biopolymer ε-PLL inhibited growth of *S. epidermidis* on the membranes but did not affect the cytocompatibility to cultured human fibroblast.
BC/Ag nanoparticle composite [54,55]		Environmentally benign and facile approach; Sustained release of Ag; Prolonged antibacterial performance against *Staphylococcus aureus*.
Silymarin (SMN)-zein nanoparticle/BC nanocomposite [56]		Change of wettability and swelling; Antioxidant and antibacterial activity; Air-dried SMN-zein/BC nanocomposite slow down the lipid oxidation.
BC/Octenidin/Poloxamer hybrid system [57]	Drug deliveryWound treatment	Long term controlled release of octenidine; Improved mechanical and antimicrobial properties; Ready-to-use system with Poloxamer-loaded BC for advanced treatment of infected wounds; Non toxicity in test with shell-less hen's egg model.
BC/CMC/Methotrexate [58]		Impact of DS-CMC on methotrexate loading; Topical treatment of psoriasis; Decrease of the elastic modulus as the degree of substitution (DS) of CMC increased;
BC/PHEMA Hydrogel matrice [59]	Biomedical application	New modification: in situ ultraviolet (UV) radical polymerization; Tensile strength increased; Nontoxic; Rat mesenchymal stem cells (rMSCs) proliferation; Tissue replacement and wound healing.
BC with tuned porosity [60]	Tissue engineering	Higher pore size than native BC to allow muscle cell ingrowth; Small decrease in mechanical strength.

Table 2. *Cont.*

Modification	Application	Resulting Properties
BC/PVA composite [61] BC/Hyaluronic acid (HA) [62]	Artificial cornea	Higher visible light transmittance than plain BC.
BC/urinary bladder matrix [63]	Retinal pigment epithelium	Higher adhesion and proliferation of retinal pigment epithelium cells than uncoated BC; Closer recapitulation of the *in vivo* cell phenotype than uncoated BC.
BC/iron oxide nanoparticles [64]	Blood vessels	Introduction of magnetic domains; Young's modulus correspond to values for blood vessels.

4.1. In Situ Modifications

Several studies identify in situ modifications as straightforward approach for introduction of particular functionality to BC by addition of reinforcement material (chitosan, gelatin, poly-3-hydroxybutirate, nanomaterials, clays, silica) to the bacterial culture medium, mostly at the beginning of BC production. The great advantage of such a process is encaging materials that become part of the fibrils, thus enhancing BC by altering mainly the physical–mechanical properties of BC fibrils. Moreover, new functionalities also can be introduced. Recent work of Gao et al. [65] propose in situ introduction of glucose being pre-modified with carboxyfluorescein (6CF), which supplements the BC with green fluorescence signal based on ultraviolet (UV) spectroscopy and confocal microscopy detection as presented by Figure 5.

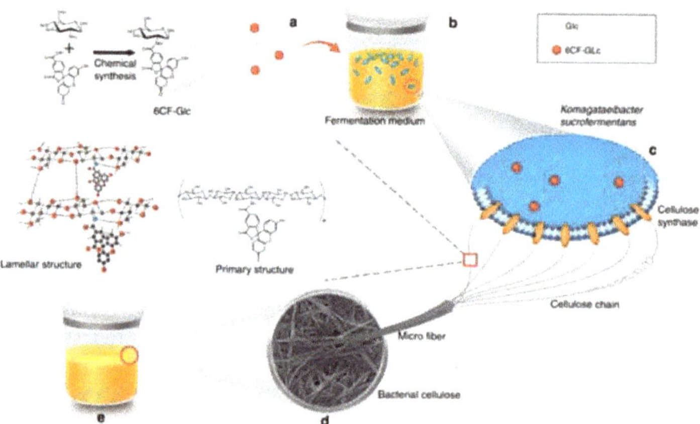

Figure 5. Synthesis of 6CF-BC by in situ microbial fermentation method, using glucose (Glc) modified with 6CF as a carbon source for *K. sucrofermentans* fermentation. (**a**) Glc and 6CF-Glc molecules; (**b**) microorganism fermentation; (**c**) the synthesis of 6CF-BC fibers through *K. sucrofermentans*, (**d**) microstructure of 6CF-BC; (**e**) the 6CF-BC pellicle obtained through microorganism fermentation. Reproduced from [65], with permission from Nature Communications, 2019.

For application in regenerative medicine and tissue engineering, the BC modification emphasis is on extracellular matrix (ECM) recapitulation [66], yet approaches are application-dependent and vastly diverse. Bone tissue engineering requires the presence of a bioactive component like hydroxyapatite $Ca_5(PO_4)_3OH$ (HAp) and tricalcium phosphate (TCP) $Ca_3(PO_4)_2$ and several research works report on their inclusion within BC culture medium, resulting in BC/hyaluronic acid (HA) composite with high bone regeneration capacity. BC/HA composite prepared in the process of the cellulose biosynthesis with the introduction of aqueous HAp suspension, allows simultaneous formation of microfibrillar stripes and partial texturing of HA crystals onto them [67]. The addition of CMC in growth media modify medium's viscosity and thus positively impacts assembling of calcium-deficient Hap powders formation in post synthetic stage, while not affecting the composite biocompatibility [68]. For vascular tissue engineering applications, the heparin-modified BC was produced by adding heparin to growth media of BC-producers, thus resulting in anticoagulant sulfate groups-enriched BC-heparin hybrid [69]. Other study introduces chitosan to BC trough in situ approach, being further ex situ modified with heparin, ending up with BC/chitosan/heparin composites with antimicrobial and anticoagulant properties [70]. For tissue-regeneration procedures, where porosity is an essential property, the paraffin microspheres were added to BC culture medium, resulting in microporous BC for bone regeneration [71], urinary conduit formation [72], etc. For wound-healing and temporary artificial skin applications, the BC culturing media is supplemented with glucose, dextrin [73], potato starch, cotton gauze, *Aloe vera*, which allows processing of composites, where only morphologies and physical properties are altered and not the chemical composition of BC itself. Addition of deacetylated chitin nanocrystals to BC culture media resulted in composite with bactericidal activity [74], while CMC addition introduced the surface charge, effective for further conjugation to affibody ligands applicable in tubular bio-filtration of blood proteins [66].

Apart from published studies, the critical limitation of the in situ modification approach presents incorporation of reinforcement materials that also have antibacterial activity against BC strains, the insolubility of various materials in culture media, high surface tension towards hydrophobic materials, the lack of structure control of BC nanofibers, and introduction of particles with low suspension stability within BC growing media, etc.

In situ modification of BC porosity is not affected by the aforementioned limitations and several studies demonstrate facile procedure for pore size manipulation. As shown by Lu et al. [60], the addition of potato starch to culture medium increases BC viscosity by interrupting BC assembly during static culture and thus creating more free spaces within the fibrous network. Further culturing of muscle cells onto loose surface of produced scaffolds results in new biomaterials for hollow organ reconstruction. The procedure for processing of macro-porous and foam-like BC was recently reported by Rühs et al. [75]; they cultured *K. xylinus* in mannitol-based media by foaming and then stabilized the product with surfactant Cremodan and viscosified with xanthan to prevent water drainage (Figure 6).

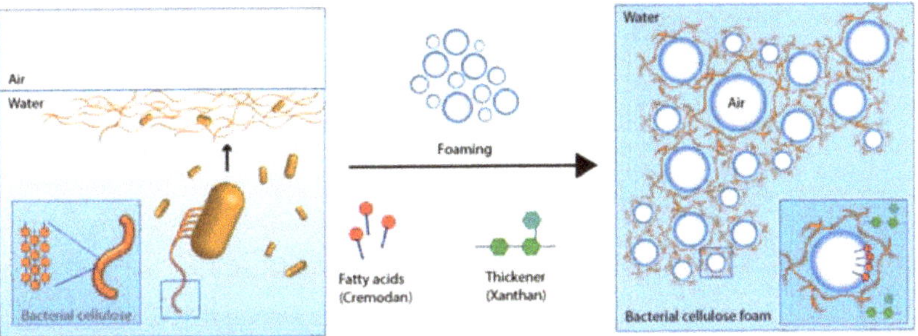

Figure 6. Schematic presentation of the BC foam formation process by *K. xylinus* suspension foaming and stabilization by Cremodan and xanthan as a thickener. Reproduced from [75], with permission from npj Biofilms and Microbiomes, 2018.

4.2. Ex Situ Modifications

Ex situ modifications are either chemical (e.g. periodate oxidation and grafting [76] or crosslinking reactions) or physical (physical absorption from solutions or particle suspensions, the homogenization or dissolving of BC mixing with additive material [77]). The BC is compounded with bioactive materials for applications such as tracking of tumor cells behavior [78], enhancement of osteoblasts cell growth in bone regeneration, fibroblast/endothelial cells guide in wound healing, etc. For the replacement of small blood vessels and improvement of the adhesion of human endothelial cells, the BC surface was modified with Arg-Gly-Asp (RGD) tripeptide, directly [79] or indirectly through xyloglucan-Gly-Arg-Gly-As-Ser (XG-GRGDS) conjugates [80]. For blood clothes control, the isolated BC from nata di coco was compounded with different fractions of kaolin [81]. To mimic the glycosaminoglycans of cartilage tissue, the surface charge was added to BC by means of chemical phosphorylation and sulfatation [82]. Incorporation of N-containing groups on BC was succeed by nitrogen plasma treatment, which also improved its porosity and enhanced the attachment of neuroblastoma (N1E-115) and human dermal microvascular endothelium (HMEC-1) cells. For application as a wound dressing, the BC was immersed into chitosan solution, forming BC/chitosan composite with high water-retention capacity [83]. For cardiovascular soft tissue replacement applications, the BC suspension was mixed with PVA, which improves the final mechanical performance [84]. Soaking of BC in silk fibroin solution results in nanocomposite with enhanced cell permissiveness, keeping the non-cytotoxicity and non-genotoxicity as in native BC [76]. For introducing antimicrobial activity against *Escherichia coli*, *Staphylococcus aureus* and *Candida albicans* while keeping biocompatibility of BC towards human embryonic kidney cells, the sodium alginate solution with silver sulfadiazine was mixed with BC slurry and further cross-linked with $CaCl_2$ [85]. Different type of nanoparticles were simultaneously formed and introduced into BC- the antimicrobial ZnO [86] and Ag nanoparticles [87], where BC was initially impregnated with zinc acetate and silver nitrate, respectively. The bone morphogenetic protein-2 was introduced into BC to promote the bone regeneration [88]. Other reported BC modifications are gentamicin-, RGD-grafted BC [89], the gelatin-grafted BC using procyanidin [90], phosphorylation [91], etc. The periodate oxidation was used for region-selective oxidation of BC and further coupling with gelatin biopolymer (Figure 7). Such composite demonstrate improved physiological degradation (compared to non-degradable, native BC) as well as capacity for accommodation of flake-like apatite minerals in short-term incubation within supersaturated simulated body fluid (SBF) [92].

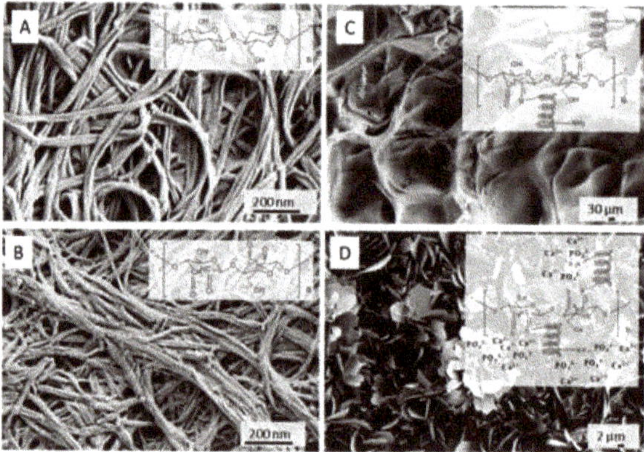

Figure 7. Scanning electron microscopy images of (**A**) native and post-synthetically modified BC; (**B**) oxidation with $NaIO_4$; (**C**) further coupling with gelatin (GEL), carbodiimide crosslinking and freeze-thawing; (**D**) in situ mineralization by incubation in (10× concentrated) simulated body fluid medium. Adapted from [92], with permission from Nanomaterials, 2019.

5. BC in Regenerative Medicine

Nanocellulose materials attract significant attention in biomedical materials research [93–95] devoted to tissue engineering [96], cell [97] and gene therapy [98], diagnostic [99] and controlled delivery [100], mainly related to their nano-features and properties arising from them. For BC, there is also ultra-high purity and net-like morphology similar to (human) collagen as a biomimetic feature, which facilitates applications such as artificial skin (Figure 8a), vascular grafts (Figure 8b), tissue-engineering scaffolds, dental implants, medical pads, artificial bone and cartilage, delivery of drugs, proteins and hormones [101]. Several commercially available products are available on market, applied during skin transplantation, second and third degree ulcer treatment, decubitus, substitution of dura mater in bran, recovery of periodontal tissues, etc. The biocompatibility assessment of BC implant, by means of chronic inflammation, foreign body responses, cell ingrowth, and angiogenesis evidence no macroscopic signs of inflammation around the implants, absence of fibrotic capsule or giant cells and fibroblasts infiltration without chronic inflammatory reaction [102].

BC efficiency in wound healing generally relies on effective cohesion with wound boundaries, preservation of a moist environment (important for re-epithelization) combined with exudates retention capacity, high mechanical strength at wet state, liquid/gasses permeability, very low risk for irritation due to its ultra-high purity, and ease of wound inspection due to its transparency [103–105], etc. In case of chronic wound treatment with BC-based wound dressing materials, the reduction of proteolytic enzymes activity, cytokines and production of reactive oxygen species are reported.

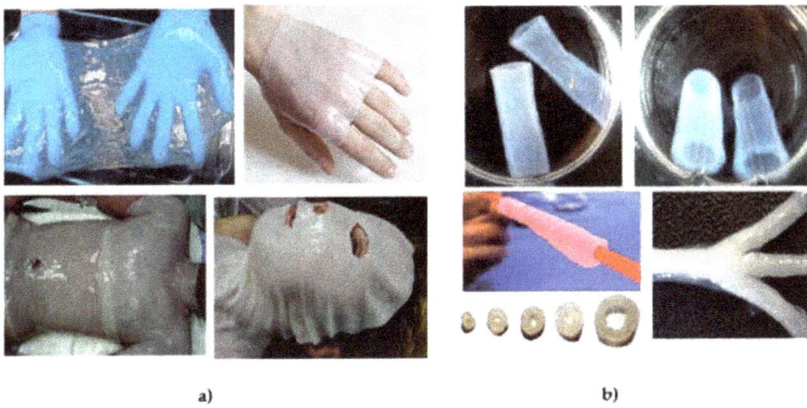

Figure 8. (a) BC dressings as produced and when applied on wounded torso, face and hand. Reproduced from [106], with permission from Biomacromolecules, 2007; (b) vascular graft and blood vessel tubes with different sizes and shape, produced by fermentation onto a branched silicone tube. Reproduced from [107–109], with permission from Frontiers, 2016, European Polymer Journal, 2014 and Biotechnology and Bioengeneering, 2007, respectively.

Even though BC possess many insintric features that encourage its use in wound dressing, its commercial dissemination is not exhaustively exploited yet [104]. The first BC-based commercial medical product was Biofill®, a thin BC film with a water content of 8.5%. Material is used as a temporary skin substitute and wound dressing in treatment of basal cell carcinoma, severe burns, dermal abrasions, chronic ulcers as well as at donor and receptor sites in skin grafts. Pain relief, close adhesion to the wound bed, spontaneous detachment following reepithelization and reduced treatment times as well as costs, yet limited elasticity, when applied in areas of great mobility, are related to this product [110]. Membracell® is also a temporary skin substitute used in treatment of burns and ulcers, sim providing pain relief, reduced infection, faster healing, etc. Bionext® and Xcell® are wound-dressing materials with similar outcomes [111]. Nanoderm™ is wound treatment product for acute and chronic wounds, allowing a barrier to infections while allowing gaseous exchange, exudate evaporation, and pain alleviation, acting as a regenerative tissue scaffold to affect fibroblast, endothelial and keratinocyte function, enhancing granulation tissue formation and epithelization [112]. The Cellumed® product is used in veterinary medicine for treatment of large surface wounds on horses and [113].

Further incorporation of inorganic (Ag [114], ZnO [114], CuO [115] and TiO_2 particles [116]) and organic antimicrobial agents (lysozyme [117], ε-poly lysine [53], nisin [118] garlics' allicin [119]), evoke their effectiveness against several bacterial strains (*Staphylococcus aureus* and *Escherichia coli*), as well as fungal strains (*Aspergilus niger* and *Candida albicans*). Abdominal hernia treatment is another application of BC as a dressing material, where better absorption in native tissue with less risk of mesh-related infections, impact and hypersensitivity at the implant site were reported [120].

A recent strategy in treatment of skin injuries is incorporation of mesenchymal stem cells, the adult pluripotent cells that can differentiate more than two cell times [121]. Loh et al. [122] seeded the human epidermal keratinocytes and dermal fibroblasts onto BC/acrylic acid hydrogel and further transferred them to a wound, reporting that the procedure accelerated the healing process.

Porosity, mouldability, foldability, hemocompatibility and good mechanical properties are attributes which position BC also in blood vessel replacement applications [123]. Especially in replacement of small blood vessels (<5 mm) as alternative to thorax or legs-harvested vessels or synthetic Dacron, extended Polytetrafluoroethylene (ePTFE) and polyurethane (PU) materials [124]. Control over porosity is prime requirement, as proliferation and migration of endothelial cells within the membrane is essential when semi-synthetic products are considered. Composite with graphene oxide [125], functionalization with chimeric proteins (conjugates of cellulose binding module and RGD adhesion peptides) [126], blending with PVA polymer [127] are among reported studies where coagulation issues and hemocompatibility are toughly investigated.

A commercial product used in the area of guided tissue and bone regeneration is Gengiflex®, the two-layer membrane comprised of native and alkali-modified BC, used for treating the osseous deficiency surround TiAl$_6$V$_4$ (IMZ) dental implant with simultaneous restoration of the aesthetic and mouth function [128]. This product was shown to support recovery of periodontal tissue by reduced inflammatory response, requiring fewer surgical steps. Saska et al. [129] reported a combination of glycine-modified BC and type I collagen with high alkaline phosphatase (ALP) activity for bone tissue regeneration. For same application, the hydroxyapatite-coated BC was investigated by Ahn et al. and new bone formation within rat calvarian defect model in 8 weeks study was defined as highly promising outcome [130]. Complexation capacity of phosphorylated BC towards calcium was utilized in study of augmentation of mineralization yields and migration of bone-forming osteoprogenitor cells [131,132].

In a recent study, Gorgieva et al. [92] combined BC membrane with gelatin utilizing successive periodate oxidation and a freeze-thawing/carbodiimide crosslinking procedure, which forms μ-porous composite membrane. Acting as a barrier for fibroblast penetration, the membrane did not evoke any cytotoxic effects toward human fibroblast (MRC-5) cells, while the same preferentially attached on a gelatin porous site (Figure 9).

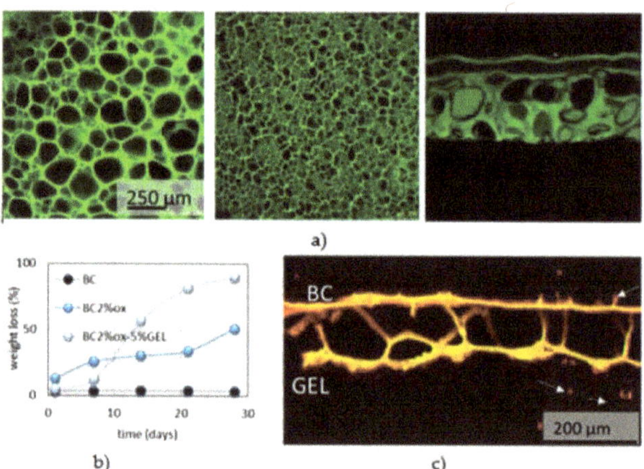

Figure 9. (a) Fluorescent microscopy images of top, bottom and cross-section aspect of BC-gelatin composite membranes; (b) their degradation kinetic; (c) barrier effect towards MRC-5 cells. Adapted from [92], with permission from Nanomaterials, 2019.

In neural tissue engineering, Innala et al. [133] reported that BC adapts to the SH-SY5Y neuroblastoma cells, which adhered, proliferated and differentiated towards mature neurons as measured by electrophysiological data. A study generated a 3D model that can be used for developing in vitro disease models. For example, combining this scaffold with human-induced pluripotent stem cells that have been derived from diseased patients, the 3D model can be used for detailed investigations of neurodegenerative disease mechanisms and in the search for new therapeutics [133].

The absence of suitable polymers and proteins, and the presence of low endotoxin units (according to the U.S. Food and Drug Administration (FDA) legislation), further expands the BC application portfolio towards drug-delivery applications [134,135], especially to tuning the drug release kinetic and optimization of drug concentration. Amin et al. [136] reported pH sensitive hydrogel formulations of BC with polyacrylic acid and bovine serum albumin as a model drug. Another study investigate BC membranes with added photosensitizer, chloroaluminum phthalocyanine for photodynamic therapy in skin cancer treatment [137].

Other cellulosic fibers (nitrocellulose in particular) have a long history as anchoring substrate for antibody conjugation in diagnostic assays [138], where also the BC appear as suitable candidate. Major effort in this area is given to processing on homogenous, 3D films in order to increase the quantity of antibodies to be further anchored. BC combination with PVA was investigated as artificial cornea [61] and aortic heart valve leaflet [139]. Tronser et al. [140] identify BC as convenient material enabling for long-term maintenance of mouse embryonic stem cells, simultaneously facilitating their culturing and handling.

6. Perspectives and Challenges for BC

BC offers an inestimable platform for development within the biomedical field, especially towards high-tech products, from nursing and diagnostic to theranostic and highly demanding regenerative, tissue-engineering products. However, more effort needs to be made in initial production steps and the fact that AAB productivity towards BC production varies strongly among different species and strains, as well as the carbon source, opens room for additional basic research input in this area. Traditional carbon sources in BC production are glucose, fructose and glycerol, which significantly increases expenses, presenting ~30% of total BC production costs. The industrial wastes or by-products have recently been proposed as cheap local sources for BC production. Some examples are corn steep liquor (CSL)-fructose medium, which is a fully enriched medium with minerals, inositol, nicotinic acid, thiamine and pantothenic acid. Date syrup and molasses are other alternatives, being highly competitive with traditional Hestrin–Schramm and Yamanaka media in BC production. Alternative carbon sources, (i.e., what straw, fruit juices, rotten fruit, waste from cotton textiles, dairy industries, biodiesel industries are already suggested) may potentially enlarge, speed up and cheapen BC production. As such, they have not been fully explored to the stage of semi-final, biomedical products. This in turn will seek more facile, cost-effective and industry-translatable modifications beyond standard post-synthetic oxidation and grafting pathways. Potential "housing" of selected and suitable biopolymers or particulates within BC during the synthetic procedure while keeping in mind no restricted BC production is one way to tackle the problem.

Author Contributions: Conceptualization: S.G. and J.T.; investigation: S.G. and J.T.; resources: S.G. and J.T.; writing—original draft preparation: S.G. and J.T.; writing—review and editing: S.G. and J.T.; visualization: S.G. and J.T.; supervision: S.G.; project administration: S.G.; funding acquisition: S.G.

Funding: The authors acknowledge the project (Z7-7169) and programs Textile chemistry (P2-0118 (B)) and Physico-Chemical Processes on the Surface Layers and Application of Nanoparticles (P2-0006), under financial support from Slovenian Research Agency.

Acknowledgments: Silvo Hribernik for SEM images and Nina Jančič for photography of BC product.

References

1. Krasteva, P.V.; Bernal-Bayard, J.; Travier, L.; Martin, F.A.; Kaminski, P.A.; Karimova, G.; Fronzes, R.; Ghigo, J.M. Insights into the structure and assembly of a bacterial cellulose secretion system. *Nat. Commun.* **2017**, *8*, 25–28. [CrossRef] [PubMed]
2. Wang, S.S.; Han, Y.H.; Chen, J.L.; Zhang, D.C.; Shi, X.X.; Ye, Y.X.; Chen, D.L.; Li, M. Insights into bacterial cellulose biosynthesis from different carbon sources and the associated biochemical transformation pathways in *Komagataeibacter* sp. W1. *Polymers* **2018**, *9*, 963. [CrossRef] [PubMed]
3. Trček, J.; Barja, F. Updates on quick identification of acetic acid bacteria with a focus on the 16S–23S rRNA gene internal transcribed spacer and the analysis of cell proteins by MALDI-TOF mass spectrometry. *Int. J. Food Microbiol.* **2015**, *196*, 137–144. [CrossRef] [PubMed]
4. Škraban, J.; Cleenwerck, I.; Vandamme, P.; Fanedl, L.; Trček, J. Genome sequences and description of novel exopolysaccharides producing species *Komagataeibacter pomaceti* sp. nov. and reclassification of *Komagataeibacter kombuchae* (Dutta and Gachhui 2007) Yamada et al. 2013 as a later heterotypic synonym of *Komagataeibacter*. *Syst. Appl. Microbiol.* **2018**, *41*, 581–592. [PubMed]
5. Slapšak, N.; Cleenwerck, I.; de Vos, P.; Trček, J. *Gluconacetobacter maltaceti* sp. nov., a novel vinegar producing acetic acid bacterium. *Syst. Appl. Microbiol.* **2013**, *36*, 17–21. [CrossRef] [PubMed]
6. Castro, C.; Zuluaga, R.; Álvarez, C.; Putaux, J.-L.; Caro, G.; Rojas, O.J.; Mondragon, I.; Gañán, P. Bacterial cellulose produced by a new acid-resistant strain of *Gluconacetobacter* genus. *Carbohydr. Polym.* **2012**, *89*, 1033–1037. [CrossRef]
7. Castro, C.; Cleenwerck, I.; Trcek, J.; Zuluaga, R.; De Vos, P.; Caro, G.; Aguirre, R.; Putaux, J.L.; Gañán, P. *Gluconacetobacter medellinensis* sp. nov., cellulose- and non-cellulose-producing acetic acid bacteria isolated from vinegar. *Int. J. Syst. Evol. Microbiol.* **2013**, *63*, 1119–1125. [CrossRef]
8. Zhang, H.; Chen, C.; Zhu, C.; Sun, D. Production of Bacterial Cellulose by *Acetobacter Xylinum*: Effects of Carbon/Nitrogen-ratio on Cell Growth and Metabolite Production. *Cellulose Chem. Technol.* **2016**, *50*, 997–1003.
9. Li, J.; Chen, G.; Zhang, R.; Wu, H.; Zeng, W.; Liang, Z. Production of high crystallinity type-I cellulose from *Komagataeibacter hansenii* JR-02 isolated from Kombucha tea. *Biotechnol. Appl. Biochem.* **2019**, *66*, 108–118. [CrossRef]
10. Morgan, J.L.W.; Strumillo, J.; Zimmer, J. Crystallographic snapshot of cellulose synthesis and membrane translocation. *Nature* **2012**, *493*, 181–186. [CrossRef]
11. Ross, P.; Mayer, R.; Benziman, M. Cellulose biosynthesis and function in bacteria. *Microbiol. Rev.* **1991**, *55*, 35–58. [PubMed]
12. McNamara, J.T.; Morgan, J.L.W.; Zimmer, J. A Molecular Description of Cellulose Biosynthesis. *Annu. Rev. Biochem.* **2015**, *84*, 895–921. [CrossRef] [PubMed]
13. Römling, U.; Galperin, M.Y. Bacterial cellulose biosynthesis: Diversity of operons, subunits, products, and functions. *Trends Microbiol.* **2015**, *23*, 545–557. [CrossRef] [PubMed]
14. Yang, H.; McManus, J.; Oehme, D.; Singh, A.; Yingling, Y.G.; Tien, M.; Kubicki, J.D. Simulations of Cellulose Synthesis Initiation and Termination in Bacteria. *J. Phys. Chem. B* **2019**, *17*, 3699–3705. [CrossRef] [PubMed]
15. Ryngajłło, M.; Kubiak, K.; Jędrzejczak-Krzepkowska, M.; Jacek, P.; Bielecki, S. Comparative genomics of the *Komagataeibacter* strains—Efficient bionanocellulose producers. *Microbiologyopen* **2018**, *8*, 1–25. [CrossRef] [PubMed]
16. Jakob, F.; Pfaff, A.; Novoa-Carballal, R.; Rübsam, H.; Becker, T.; Vogel, R.F. Structural analysis of fructans produced by acetic acid bacteria reveals a relation to hydrocolloid function. *Carbohydr. Polym.* **2013**, *92*, 1234–1242. [CrossRef] [PubMed]
17. Brandt, J.U.; Jakob, F.; Behr, J.; Geissler, A.J.; Vogel, R.F. Dissection of exopolysaccharide biosynthesis in *Kozakia baliensis*. *Microb. Cell Fact.* **2016**, *15*, 1–13. [CrossRef]
18. Morris, V.J.; Brownsey, G.J.; Cairns, P.; Chilvers, G.R.; Miles, M.J. Molecular origins of acetan solution properties. *Int. J. Biol. Macromol.* **1989**, *11*, 326–328. [CrossRef]
19. Ishida, T.; Sugano, Y.; Nakai, T.; Shoda, M. Effects of Acetan on Production of Bacterial Cellulose by *Acetobacter xylinum*. *Biosci. Biotechnol. Biochem.* **2003**, *66*, 1677–1681. [CrossRef]
20. Schramm, B.M.; Hestrin, S. Factors affecting Production of Cellulose at the Air/Liquid Interface of a Culture of *Acetobacter xylinum*. *J. Gen. Microbiol.* **1954**, *11*, 123–129. [CrossRef]

21. Naritomi, T.; Kouda, T.; Yano, H.; Yoshinaga, F. Effect of ethanol on bacterial cellulose production from fructose in continuous culture. *J. Ferment. Bioeng.* **1998**, *85*, 598–603. [CrossRef]

22. Toda, K.; Asakura, T.; Fukaya, M.; Entani, E.; Kawamura, Y. Cellulose production by acetic acid-resistant Acetobacter xylinum. *J. Ferment. Bioeng.* **1997**, *84*, 228–231. [CrossRef]

23. Lu, H.; Jia, Q.; Chen, L.; Zhang, L. Effect of Organic Acids on Bacterial Cellulose Produced by Acetobacter xylinum. *J. Microbiol. Biotechnol.* **2016**, *5*, 1–6.

24. Li, Y.; Tian, C.; Tian, H.; Zhang, J.; He, X.; Ping, W.; Lei, H. Improvement of bacterial cellulose production by manipulating the metabolic pathways in which ethanol and sodium citrate involved. *Appl. Microbiol. Biotechnol.* **2012**, *96*, 1479–1487. [CrossRef] [PubMed]

25. Lu, Z.; Zhang, Y.; Chi, Y.; Xu, N.; Yao, W.; Sun, Y. Effects of alcohols on bacterial cellulose production by Acetobacter xylinum 186. *J. Microbiol. Biotechnol.* **2011**, *27*, 2281–2285. [CrossRef]

26. Matsuoka, M.; Tsuchida, T.; Matsushita, K.; Adachi, O.; Yoshinaga, F. A Synthetic Medium for Bacterial Cellulose Production by Acetobacter xylinum subsp. Sucrofermentans. *Biosci. Biotechnol. Biochem.* **2011**, *60*, 575–579. [CrossRef]

27. Cheng, Z.; Yang, R.; Liu, X.; Liu, X.; Chen, H. Green synthesis of bacterial cellulose via acetic acid pre-hydrolysis liquor of agricultural corn stalk used as carbon source. *Bioresour. Technol.* **2017**, *234*, 8–14. [CrossRef]

28. Yang, X.-Y.; Huang, C.; Guo, H.-J.; Xiong, L.; Luo, J.; Wang, B.; Jin, X.Q.; Chen, X.F.; Chen, X.-D. Bacterial cellulose production from the litchi extract by *Gluconacetobacter xylinus*. *Prep. Biochem. Biotechnol.* **2016**, *46*, 39–43. [CrossRef]

29. Fan, X.; Gao, Y.; He, W.; Hu, H.; Tian, M.; Wang, K.; Pan, S. Production of nano bacterial cellulose from beverage industrial waste of citrus peel and pomace using *Komagataeibacter xylinus*. *Carbohydr. Polym.* **2016**, *151*, 1068–1072. [CrossRef]

30. Huang, C.; Yang, X.-Y.; Xiong, L.; Guo, H.-J.; Luo, J.; Wang, B.; Zhang, H.R.; Lin, X.Q.; Chen, X.-D. Evaluating the possibility of using acetone-butanol-ethanol (ABE) fermentation wastewater for bacterial cellulose production by *Gluconacetobacter xylinus*. *Lett. Appl. Microbiol.* **2015**, *60*, 491–496. [CrossRef]

31. Lin, D.; Lopez-Sanchez, P.; Li, R.; Li, Z. Production of bacterial cellulose by *Gluconacetobacter hansenii* CGMCC 3917 using only waste beer yeast as nutrient source. *Bioresour. Technol.* **2014**, *151*, 113–119. [CrossRef] [PubMed]

32. García-Lomillo, J.; González-SanJosé, M.L. Applications of Wine Pomace in the Food Industry: Approaches and Functions. *Compr. Rev. Food Sci. Food Saf.* **2017**, *16*, 3–22. [CrossRef]

33. Bayrak, E.; Büyükkileci, A.O. Utilization of white grape pomace for lactic acid production. *GIDA* **2018**, *43*, 129–138. [CrossRef]

34. Molina-Ramírez, C.; Enciso, C.; Torres-Taborda, M.; Zuluaga, R.; Gañán, P.; Rojas, O.J.; Castro, C. Effects of alternative energy sources on bacterial cellulose characteristics produced by *Komagataeibacter medellinensis*. *Int. J. Biol. Macromol.* **2018**, *117*, 735–741. [CrossRef] [PubMed]

35. Keshk, S.M.A.S. Vitamin C enhances bacterial cellulose production in *Gluconacetobacter xylinus*. *Carbohydr. Polym.* **2014**, *99*, 98–100. [CrossRef]

36. Rani, M.U.; Rastogi, N.K.; Appaiah, K.A.A. Statistical optimization of medium composition for bacterial cellulose production by *Gluconacetobacter hansenii* UAC09 using coffee cherry husk extract—An agro-industry waste. *J. Microbiol. Biotechnol.* **2011**, *21*, 739–745. [CrossRef] [PubMed]

37. Watanabe, K.; Tabuchi, M.; Morinaga, Y.; Yoshinaga, F. Structural Features and Properties of Bacterial Cellulose Produced in Agitated Culture. *Cellulose* **1998**, *5*, 187–200. [CrossRef]

38. Wang, J.; Tavakoli, J.; Tang, Y. Bacterial cellulose production, properties and applications with different culture methods—A review. *Carbohydr. Polym.* **2019**, *219*, 63–76. [CrossRef]

39. Chao, Y.P.; Sugano, Y.; Kouda, T.; Yoshinaga, F.; Shoda, M. Production of bacterial cellulose by *Acetobacter xylinum* with an air-lift reactor. *Biotechnol. Tech.* **1997**, *11*, 829–832. [CrossRef]

40. Serafica, G.; Mormino, R.; Bungay, H. Inclusion of solid particles in bacterial cellulose. *Appl. Microbiol. Biotechnol.* **2002**, *58*, 756–760.

41. Lu, H.; Jiang, X. Structure and Properties of Bacterial Cellulose Produced Using a Trickling Bed Reactor. *Appl. Biochem. Biotechnol.* **2014**, *172*, 3844–3861. [CrossRef] [PubMed]

42. Sriplai, N.; Mongkolthanaruk, W.; Eichhorn, S.J.; Pinitsoontorn, S. Magnetically responsive and flexible bacterial cellulose membranes. *Carbohydr. Polym.* **2018**, *192*, 251–262. [CrossRef] [PubMed]

43. Bäckdahl, H.; Helenius, G.; Bodin, A.; Nannmark, U.; Johansson, B.R.; Risberg, B.; Gatenholm, P. Mechanical properties of bacterial cellulose and interactions with smooth muscle cells. *Biomaterials* **2006**, *27*, 2141–2149. [CrossRef] [PubMed]

44. Keshk, S.M.A.S.; Nada, A.M.A. Heterogeneous Derivatization of Bacterial and Plant Cellulose. *Biosci. Biotechnol. Res. Asia* **2016**, *1*, 39–42.

45. Rebelo, A.R.; Archer, A.J.; Chen, X.; Liu, C.; Yang, G.; Liu, Y. Dehydration of bacterial cellulose and the water content effects on its viscoelastic and electrochemical properties. *Sci. Technol. Adv. Mater.* **2018**, *19*, 203–211. [CrossRef] [PubMed]

46. Seifert, M.; Hesse, S.; Kabrelian, V.; Klemm, D. Controlling the water content of never dried and reswollen bacterial cellulose by the addition of water-soluble polymers to the culture medium. *J. Polym. Sci. Part A Polym. Chem.* **2004**, *42*, 463–470. [CrossRef]

47. Bottan, S.; Robotti, F.; Jayathissa, P.; Hegglin, A.; Bahamonde, N.; Heredia-Guerrero, J.A.; Bayer, I.S.; Scarpellini, A.; Merker, H.; Lindebnblatt, N.; et al. Surface-structured bacterial cellulose with guided assembly-based biolithography (GAB). *ACS Nano* **2015**, *9*, 206–219. [CrossRef]

48. Wu, H.; Williams, G.R.; Wu, J.; Wu, J.; Niu, S.; Li, H.; Wang, H.; Zhu, L. Regenerated chitin fibers reinforced with bacterial cellulose nanocrystals as suture biomaterials. *Carbohydr. Polym.* **2018**, *180*, 304–313. [CrossRef]

49. Qiu, Y.; Qiu, L.; Cui, J.; Wei, Q. Bacterial cellulose and bacterial cellulose-vaccarin membranes for wound healing. *Mater. Sci. Eng. C* **2016**, *59*, 303–309. [CrossRef]

50. Wu, C.-N.; Fuh, S.-C.; Lin, S.-P.; Lin, Y.-Y.; Chen, H.-Y.; Liu, J.-M.; Cheng, K.-C. TEMPO-Oxidized Bacterial Cellulose Pellicle with Silver Nanoparticles for Wound Dressing. *Biomacromolecules* **2018**, *19*, 544–554. [CrossRef]

51. Khalid, A.; Khan, R.; Ul-Islam, M.; Khan, T.; Wahid, F. Bacterial cellulose-zinc oxide nanocomposites as a novel dressing system for burn wounds. *Carbohydr. Polym.* **2017**, *164*, 214–221. [CrossRef] [PubMed]

52. Khalid, A.; Ullah, H.; Ul-Islam, M.; Khan, R.; Khan, S.; Ahmad, F.; Khan, T.; Wahid, F. Bacterial cellulose–TiO$_2$ nanocomposites promote healing and tissue regeneration in burn mice model. *RSC Adv.* **2017**, *7*, 47662–47668. [CrossRef]

53. Fürsatz, M.; Skog, M.; Sivlér, P.; Palm, E.; Aronsson, C.; Skallberg, A.; Greczynski, G.; Khalaf, H.; Bengtsson, T.; Aili, D. Functionalization of bacterial cellulose wound dressings with the antimicrobial peptide ε-poly-L-Lysine. *Biomed. Mater.* **2018**, *13*, 025014. [CrossRef] [PubMed]

54. Yang, G.; Xie, J.; Deng, Y.; Bian, Y.; Hong, F. Hydrothermal synthesis of bacterial cellulose/AgNPs composite: A 'green' route for antibacterial application. *Carbohydr. Polym.* **2012**, *87*, 2482–2487. [CrossRef]

55. Maneerung, T.; Tokura, S.; Rujiravanit, R. Impregnation of silver nanoparticles into bacterial cellulose for antimicrobial wound dressing. *Carbohydr. Polym.* **2008**, *72*, 43–51. [CrossRef]

56. Tsai, Y.-H.; Yang, Y.-N.; Ho, Y.-C.; Tsai, M.-L.; Mi, F.-L. Drug release and antioxidant/antibacterial activities of silymarin-zein nanoparticle/bacterial cellulose nanofiber composite films. *Carbohydr. Polym.* **2018**, *180*, 286–296. [CrossRef] [PubMed]

57. Alkhatib, Y.; Dewaldt, M.; Moritz, S.; Nitzsche, R.; Kralisch, D.; Fischer, D. Controlled extended octenidine release from a bacterial nanocellulose/Poloxamer hybrid system. *Eur. J. Pharm. Biopharm.* **2017**, *112*, 164–176. [CrossRef]

58. de Lima Fontes, M.; Meneguin, A.B.; Tercjak, A.; Gutierrez, J.; Cury, B.S.F.; dos Santos, A.M.; Ribeiro, S.J.L.; Barud, H.S. Effect of in situ modification of bacterial cellulose with carboxymethylcellulose on its nano/microstructure and methotrexate release properties. *Carbohydr. Polym.* **2018**, *179*, 126–134. [CrossRef] [PubMed]

59. Hobzova, R.; Hrib, J.; Sirc, J.; Karpushkin, E.; Michalek, J.; Janouskova, O.; Gatenholm, P. Embedding of Bacterial Cellulose Nanofibers within PHEMA Hydrogel Matrices: Tunable Stiffness Composites with Potential for Biomedical Applications. *J. Nanomater.* **2018**, *2018*, 1–11. [CrossRef]

60. Lv, X.; Yang, J.; Feng, C.; Li, Z.; Chen, S.; Xie, M.; Xu, Y. Bacterial Cellulose-Based Biomimetic Nanofibrous Scaffold with Muscle Cells for Hollow Organ Tissue Engineering. *ACS Biomater. Sci. Eng.* **2016**, *2*, 19–29. [CrossRef]

61. Wang, J.; Gao, C.; Zhang, Y.; Wan, Y. Preparation and in vitro characterization of BC/PVA hydrogel composite for its potential use as artificial cornea biomaterial. *Mater. Sci. Eng. C* **2010**, *30*, 214–218. [CrossRef]

62. Jia, Y.; Zhu, W.; Zheng, M.; Huo, M.; Zhong, C. Bacterial cellulose/hyaluronic acid composite hydrogels with improved viscoelastic properties and good thermodynamic stability. *Plast. Rubber Compos.* **2018**, *47*, 165–175. [CrossRef]

63. Gonçalves, S.; Rodrigues, I.P.; Padrão, J.; Silva, J.P.; Sencadas, V.; Lanceros-Mendez, S.; Rodrigues, L.R. Acetylated bacterial cellulose coated with urinary bladder matrix as a substrate for retinal pigment epithelium. *Colloids Surf. B* **2016**, *139*, 1–9. [CrossRef] [PubMed]

64. Arias, S.L.; Shetty, A.R.; Senpan, A.; Echeverry-Rendón, M.; Reece, L.M.; Allain, J.P. Fabrication of a Functionalized Magnetic Bacterial Nanocellulose with Iron Oxide Nanoparticles. *J. Vis. Exp.* **2016**. [CrossRef] [PubMed]

65. Gao, M.; Li, J.; Bao, Z.; Hu, M.; Nian, R.; Feng, D.; Zhang, H. A natural in situ fabrication method of functional bacterial cellulose using a microorganism. *Nat. Commun.* **2019**, *10*, 437. [CrossRef]

66. Orelma, H.; Morales, L.O.; Johansson, L.S.; Hoeger, I.C.; Filpponen, I.; Castro, C.; Laine, J. Affibody conjugation onto bacterial cellulose tubes and bioseparation of human serum albumin. *RSC Adv.* **2014**, *4*, 51440–51450. [CrossRef]

67. Romanov, D.P.; Khripunov, A.K.; Baklagina, Y.G.; Severin, A.V.; Lukasheva, N.V.; Tolmachev, D.A.; Klechkovskaya, V.V. Nanotextures of composites based on the interaction between hydroxyapatite and cellulose *Gluconacetobacter xylinus*. *Glas. Phys. Chem.* **2014**, *40*, 367–374. [CrossRef]

68. Grande, C.J.; Torres, F.G.; Gomez, C.M.; Bañó, M.C. Nanocomposites of bacterial cellulose/hydroxyapatite for biomedical applications. *Acta Biomater.* **2009**, *5*, 1605–1615. [CrossRef]

69. Wan, Y.; Gao, C.; Han, M.; Liang, H.; Ren, K.; Wang, Y.; Luo, H. Preparation and characterization of bacterial cellulose/heparin hybrid nanofiber for potential vascular tissue engineering scaffolds. *Polym. Adv. Technol.* **2011**, *22*, 2643–2648. [CrossRef]

70. Wang, J.; Wan, Y.; Huang, Y. Immobilisation of heparin on bacterial cellulose-chitosan nano-fibres surfaces via the cross-linking technique. *IET Nanobiotechnol.* **2012**, *6*, 52. [CrossRef]

71. Brackmann, C.; Zaborowska, M.; Sundberg, J.; Gatenholm, P.; Enejder, A. In situ imaging of collagen synthesis by osteoprogenitor cells in microporous bacterial cellulose scaffolds. *Tissue Eng. Part C Methods* **2012**, *18*, 227–234. [CrossRef] [PubMed]

72. Bodin, A.; Bharadwaj, S.; Wu, S.; Gatenholm, P.; Atala, A.; Zhang, Y. Tissue-engineered conduit using urine-derived stem cells seeded bacterial cellulose polymer in urinary reconstruction and diversion. *Biomaterials* **2010**, *31*, 8889–8901. [CrossRef] [PubMed]

73. Stumpf, T.R.; Pértile, R.A.N.; Rambo, C.R.; Porto, L.M. Enriched glucose and dextrin mannitol-based media modulates fibroblast behavior on bacterial cellulose membranes. *Mater. Sci. Eng. C* **2013**, *33*, 4739–4745. [CrossRef] [PubMed]

74. Butchosa, N.; Brown, C.; Larsson, P.T.; Berglund, L.A.; Bulone, V.; Zhou, Q. Nanocomposites of bacterial cellulose nanofibers and chitin nanocrystals: Fabrication, characterization and bactericidal activity. *Green Chem.* **2013**, *15*, 3404. [CrossRef]

75. Rühs, P.A.; Storz, F.; Gómez, Y.A.L.; Haug, M.; Fischer, P. 3D bacterial cellulose biofilms formed by foam templating. *NPJ Biofilms Microbiomes* **2018**, *4*, 21. [CrossRef] [PubMed]

76. Oliveira Barud, H.G.; Barud, H.d.S.; Cavicchioli, M.; do Amaral, T.S.; Junior, O.B.d.O.; Santos, D.M.; Ribeiro, S.J.L. Preparation and characterization of a bacterial cellulose/silk fibroin sponge scaffold for tissue regeneration. *Carbohydr. Polym.* **2015**, *128*, 41–51. [CrossRef] [PubMed]

77. Cai, Z.; Kim, J. Preparation and Characterization of Novel Bacterial Cellulose/Gelatin Scaffold for Tissue Regeneration Using Bacterial Cellulose Hydrogel. *J. Nanotechnol. Eng. Med.* **2010**, *1*, 021002. [CrossRef]

78. Wang, J.; Zhao, L.; Zhang, A.; Huang, Y.; Tavakoli, J.; Tang, Y. Novel Bacterial Cellulose/Gelatin Hydrogels as 3D Scaffolds for Tumor Cell Culture. *Polymers* **2018**, *10*, 581. [CrossRef]

79. Andrade, F.K.; Silva, J.P.; Carvalho, M.; Castanheira, E.M.S.; Soares, R.; Gama, M. Studies on the hemocompatibility of bacterial cellulose. *J. Biomed. Mater. Res. Part A* **2011**, *98A*, 554–566. [CrossRef]

80. Bodin, A.; Ahrenstedt, L.; Fink, H.; Brumer, H.; Risberg, B.; Gatenholm, P. Modification of Nanocellulose with a Xyloglucan–RGD Conjugate Enhances Adhesion and Proliferation of Endothelial Cells: Implications for Tissue Engineering. *Biomacromolecules* **2007**, *8*, 3697–3704. [CrossRef]

81. Véliz, D.S.; Alam, C.; Toivola, D.M.; Toivakka, M.; Alam, P. On the non-linear attachment characteristics of blood to bacterial cellulose/kaolin biomaterials. *Colloids Surf. B* **2014**, *116*, 176–182. [CrossRef] [PubMed]

82. Svensson, A.; Nicklasson, E.; Harrah, T.; Panilaitis, B.; Kaplan, D.L.; Brittberg, M.; Gatenholm, P. Bacterial cellulose as a potential scaffold for tissue engineering of cartilage. *Biomaterials* **2005**, *26*, 419–431. [CrossRef] [PubMed]

83. Lin, W.-C.; Lien, C.-C.; Yeh, H.-J.; Yu, C.-M.; Hsu, S. Bacterial cellulose and bacterial cellulose–chitosan membranes for wound dressing applications. *Carbohydr. Polym.* **2013**, *94*, 603–611. [CrossRef] [PubMed]

84. Millon, L.E.; Wan, W.K. The polyvinyl alcohol–bacterial cellulose system as a new nanocomposite for biomedical applications. *J. Biomed. Mater. Res. Part. B Appl. Biomater.* **2006**, *79B*, 245–253. [CrossRef] [PubMed]

85. Shao, W.; Liu, H.; Liu, X.; Wang, S.; Wu, J.; Zhang, R.; Huang, M. Development of silver sulfadiazine loaded bacterial cellulose/sodium alginate composite films with enhanced antibacterial property. *Carbohydr. Polym.* **2015**, *132*, 351–358. [CrossRef] [PubMed]

86. Katepetch, C.; Rujiravanit, R.; Tamura, H. Formation of nanocrystalline ZnO particles into bacterial cellulose pellicle by ultrasonic-assisted in situ synthesis. *Cellulose* **2013**, *20*, 1275–1292. [CrossRef]

87. Yang, G.; Xie, J.; Hong, F.; Cao, Z.; Yang, X. Antimicrobial activity of silver nanoparticle impregnated bacterial cellulose membrane: Effect of fermentation carbon sources of bacterial cellulose. *Carbohydr. Polym.* **2012**, *87*, 839–845. [CrossRef]

88. Shi, Q.; Li, Y.; Sun, J.; Zhang, H.; Chen, L.; Chen, B.; Wang, Z. The osteogenesis of bacterial cellulose scaffold loaded with bone morphogenetic protein-2. *Biomaterials* **2012**, *33*, 6644–6649. [CrossRef]

89. Rouabhia, M.; Asselin, J.; Tazi, N.; Messaddeq, Y.; Levinson, D.; Zhang, Z. Production of Biocompatible and Antimicrobial Bacterial Cellulose Polymers Functionalized by RGDC Grafting Groups and Gentamicin. *ACS Appl. Mater. Interfaces* **2014**, *6*, 1439–1446. [CrossRef]

90. Wang, J.; Wan, Y.Z.; Luo, H.L.; Gao, C.; Huang, Y. Immobilization of gelatin on bacterial cellulose nanofibers surface via crosslinking technique. *Mater. Sci. Eng. C* **2012**, *32*, 536–541. [CrossRef]

91. Oshima, T.; Taguchi, S.; Ohe, K.; Baba, Y. Phosphorylated bacterial cellulose for adsorption of proteins. *Carbohydr. Polym.* **2011**, *83*, 953–958. [CrossRef]

92. Gorgieva, S.; Hribernik, S. Microstructured and Degradable Bacterial Cellulose–Gelatin Composite Membranes: Mineralization Aspects and Biomedical Relevance. *Nanomaterials* **2019**, *9*, 303. [CrossRef] [PubMed]

93. Gorgieva, S.; Girandon, L.; Kokol, V. Mineralization potential of cellulose-nanofibrils reinforced gelatine scaffolds for promoted calcium deposition by mesenchymal stem cells. *Mater. Sci. Eng. C* **2017**, *73*, 478–489. [CrossRef] [PubMed]

94. Gorgieva, S.; Vivod, V.; Maver, U.; Gradišnik, L.; Dolenšek, J.; Kokol, V. Internalization of (bis)phosphonate-modified cellulose nanocrystals by human osteoblast cells. *Cellulose* **2017**, *24*, 10. [CrossRef]

95. Napavichayanun, S.; Yamdech, R.; Aramwit, P. The safety and efficacy of bacterial nanocellulose wound dressing incorporating sericin and polyhexamethylene biguanide: In vitro, in vivo and clinical studies. *Arch. Dermatol. Res.* **2016**, *308*, 123–132. [CrossRef] [PubMed]

96. Markstedt, K.; Mantas, A.; Tournier, I.; Ávila, H.M.; Hägg, D.; Gatenholm, P. 3D Bioprinting Human Chondrocytes with Nanocellulose–Alginate Bioink for Cartilage Tissue Engineering Applications. *Biomacromolecules* **2015**, *16*, 1489–1496. [CrossRef] [PubMed]

97. Lou, Y.-R.; Kanninen, L.; Kuisma, T.; Niklander, J.; Noon, L.A.; Burks, D.; Yliperttula, M. The use of nanofibrillar cellulose hydrogel as a flexible three-dimensional model to culture human pluripotent stem cells. *Stem Cells Dev.* **2014**, *23*, 380–392. [CrossRef] [PubMed]

98. Ndong Ntoutoume, G.M.A.; Grassot, V.; Brégier, F.; Chabanais, J.; Petit, J.-M.; Granet, R.; Sol, V. PEI-cellulose nanocrystal hybrids as efficient siRNA delivery agents—Synthesis, physicochemical characterization and in vitro evaluation. *Carbohydr. Polym.* **2017**, *164*, 258–267. [CrossRef] [PubMed]

99. Edwards, J.V.; Fontenot, K.R.; Prevost, N.T.; Haldane, D.; Pircher, N.; Liebner, F.; French, A.; Condon, B.D. Protease Biosensors Based on Peptide-Nanocellulose Conjugates: From Molecular Design to Dressing Interface. *Int. J. Med. Nano Res.* **2016**, *3*, 1.

100. Barbosa, A.; Robles, E.; Ribeiro, J.; Lund, R.; Carreño, N.; Labidi, J. Cellulose Nanocrystal Membranes as Excipients for Drug Delivery Systems. *Materials* **2016**, *9*, 1002. [CrossRef]

101. Gallegos, A.M.A.; Carrera, S.H.; Parra, R.; Keshavarz, T.; Iqbal, H.M.N. Bacterial Cellulose: A Sustainable Source to Develop Value-Added Products—A Review. *BioResources* **2016**, *11*, 2. [CrossRef]

102. Helenius, G.; Bäckdahl, H.; Bodin, A.; Nannmark, U.; Gatenholm, P.; Risberg, B. In vivo biocompatibility of bacterial cellulose. *J. Biomed. Mater. Res. Part A* **2006**, *76A*, 431–438. [CrossRef] [PubMed]

103. Czaja, W.; Krystynowicz, A.; Kawecki, M.; Wysota, K.; Sakiel, S.; Wróblewski, P.; Glik, J.; Nowak, M.; Bielecki, S. Biomedical Applications of Microbial Cellulose in Burn Wound Recovery. In *Cellulose: Molecular and Structural Biology*; Springer: Dordrecht, The Netherlands, 2007; pp. 307–321.

104. Portela, R.; Leal, C.R.; Almeida, P.L.; Sobral, R.G. Bacterial cellulose: A versatile biopolymer for wound dressing applications. *Microb. Biotechnol.* **2019**, *12*, 586–610. [CrossRef] [PubMed]

105. Zhang, X.C. *Science and Principles of Biodegradable and Bioresorbable Medical Polymers: Materials and Properties*; Woodhead Publishing: Sawston, Cambridge, UK, 2016; pp. 295–316.

106. Czaja, W.K.; Young, D.J.; Kawecki, M.; Brown, R.M. The Future Prospects of Microbial Cellulose in Biomedical Applications. *Biomacromolecules* **2007**, *8*, 1–12. [CrossRef] [PubMed]

107. Lina, F.; Chandra, P.; Adrianna, M.; Wankei, W. Bacterial cellulose production using a novel microbe. *Front. Bioeng. Biotechnol.* **2016**, *4*. [CrossRef]

108. Lin, N.; Dufresne, A. Nanocellulose in biomedicine: Current status and future prospect. *Eur. Polym. J.* **2014**, *59*, 302–325. [CrossRef]

109. Bodin, A.; Bäckdahl, H.; Fink, H.; Gustafsson, L.; Risberg, B.; Gatenholm, P. Influence of cultivation conditions on mechanical and morphological properties of bacterial cellulose tubes. *Biotechnol. Bioeng.* **2007**, *97*, 425–434. [CrossRef]

110. Fontana, J.D.; De Souza, A.M.; Fontana, C.K.; Torriani, I.L.; Moreschi, J.C.; Gallotti, B.J.; Farah, L.F.X. Acetobacter cellulose pellicle as a temporary skin substitute. *Appl. Biochem. Biotechnol.* **1990**, *24–25*, 253–264. [CrossRef]

111. Picheth, G.F.; Pirich, C.L.; Sierakowski, M.R.; Woehl, M.A.; Sakakibara, C.N.; de Souza, C.F.; de Freitas, R.A. Bacterial cellulose in biomedical applications: A review. *Int. J. Biol. Macromol.* **2017**, *104*, 97–106. [CrossRef]

112. Nanoderm. Available online: http://nanoderm.ca (accessed on 6 September 2019).

113. Blackburn, R. *Biodegradable and Sustainable Fibres*; Woodhead Publishing: Cambridge, UK, 2005.

114. Barud, H.S.; Regiani, T.; Marques, R.F.C.; Lustri, W.R.; Messaddeq, Y.; Ribeiro, S.J.L. Antimicrobial Bacterial Cellulose-Silver Nanoparticles Composite Membranes. *J. Nanomater.* **2011**, *2011*, 1–8. [CrossRef]

115. Almasi, H.; Jafarzadeh, P.; Mehryar, L. Fabrication of novel nanohybrids by impregnation of CuO nanoparticles into bacterial cellulose and chitosan nanofibers: Characterization, antimicrobial and release properties. *Carbohydr. Polym.* **2018**, *186*, 273–281. [CrossRef] [PubMed]

116. Khan, S.; Ul-Islam, M.; Khattak, W.A.; Ullah, M.W.; Park, J.K. Bacterial cellulose-titanium dioxide nanocomposites: Nanostructural characteristics, antibacterial mechanism, and biocompatibility. *Cellulose* **2015**, *22*, 565–579. [CrossRef]

117. Bayazidi, P.; Almasi, H.; Asl, A.K. Immobilization of lysozyme on bacterial cellulose nanofibers: Characteristics, antimicrobial activity and morphological properties. *Int. J. Biol. Macromol.* **2018**, *107*, 2544–2551. [CrossRef] [PubMed]

118. Nguyen, V.T.; Gidley, M.J.; Dykes, G.A. Potential of a nisin-containing bacterial cellulose film to inhibit Listeria monocytogenes on processed meats. *Food Microbiol.* **2008**, *25*, 471–478. [CrossRef] [PubMed]

119. Jebali, A.; Hekmatimoghaddam, S.; Behzadi, A.; Rezapor, I.; Mohammadi, B.H.; Jasemizad, T.; Sayadi, M. Antimicrobial activity of nanocellulose conjugated with allicin and lysozyme. *Cellulose* **2013**, *20*, 2897–2907. [CrossRef]

120. JPiasecka-Zelga, P.; Szulc, J.; Wietecha, J.; Ciechańska, D. An in vivo biocompatibility study of surgical meshes made from bacterial cellulose modified with chitosan. *Int. J. Biol. Macromol.* **2018**, *116*, 1119–1127. [CrossRef] [PubMed]

121. Souza, C.M.C.O.; Mesquita, L.A.F.; Souza, D.; Irioda, A.C.; Francisco, J.C.; Souza, C.F.; Carvalho, K.A.T. Regeneration of Skin Tissue Promoted by Mesenchymal Stem Cells Seeded in Nanostructured Membrane. *Transplant. Proc.* **2014**, *46*, 1882–1886. [CrossRef]

122. Loh, E.Y.X.; Mohamad, N.; Fauzi, M.B.; Ng, M.H.; Ng, S.F.; Amin, M.C.I.M. Development of a bacterial cellulose-based hydrogel cell carrier containing keratinocytes and fibroblasts for full-thickness wound healing. *Sci. Rep.* **2018**, *8*, 2875. [CrossRef]

123. Ravi, S.; Chaikof, E.L. Biomaterials for vascular tissue engineering. *Regen. Med.* **2010**, *5*, 107–120. [CrossRef]

124. Lee, S.E.; Park, Y.S. The role of bacterial cellulose in artificial blood vessels. *Mol. Cell. Toxicol.* **2017**, *13*, 257–261. [CrossRef]

125. Zhu, W.; Li, W.; He, Y.; Duan, T. In-situ biopreparation of biocompatible bacterial cellulose/graphene oxide composites pellets. *Appl. Surf. Sci.* **2015**, *338*, 22–26. [CrossRef]

126. Andrade, F.K.; Costa, R.; Domingues, L.; Soares, R.; Gama, M. Improving bacterial cellulose for blood vessel replacement: Functionalization with a chimeric protein containing a cellulose-binding module and an adhesion peptide. *Acta Biomater.* **2010**, *6*, 4034–4041. [CrossRef]

127. Leitão, A.F.; Gupta, S.; Silva, J.P.; Reviakine, I.; Gama, M. Hemocompatibility study of a bacterial cellulose/polyvinyl alcohol nanocomposite. *Colloids Surf. B* **2012**, *111*, 493–502.

128. Novaes, A.B.J.; Novaes, A.B. Bone formation over a TiAl6V4 (IMZ) implant placed into an extraction socket in association with membrane therapy (Gengiflex). *Clin. Oral Implants Res.* **1993**, *4*, 106–110. [CrossRef] [PubMed]

129. Saska, S.; Teixeira, L.N.; Tambasco de Oliveira, P.; Minarelli Gaspar, A.M.; Lima Ribeiro, S.J.; Messaddeq, Y.; Marchetto, R. Bacterial cellulose-collagen nanocomposite for bone tissue engineering. *J. Mater. Chem.* **2012**, *22*, 22102. [CrossRef]

130. Ahn, S.-J.; Shin, Y.M.; Kim, S.E.; Jeong, S.I.; Jeong, J.-O.; Park, J.-S.; Lim, Y.-M. Characterization of hydroxyapatite-coated bacterial cellulose scaffold for bone tissue engineering. *Biotechnol. Bioprocess. Eng.* **2015**, *20*, 948–955. [CrossRef]

131. Kang, Y.J. Method of Tissue Repair Using a Composite Material. Patent Number WO2016169416A, December 2011.

132. Zaborowska, M.; Bodin, A.; Bäckdahl, H.; Popp, J.; Goldstein, A.; Gatenholm, P. Microporous bacterial cellulose as a potential scaffold for bone regeneration. *Acta Biomater.* **2010**, *6*, 2540–2547. [CrossRef]

133. Innala, M.; Riebe, I.; Kuzmenko, V.; Sundberg, J.; Gatenholm, P.; Hanse, E.; Johannesson, S. 3D Culturing and differentiation of SH-SY5Y neuroblastoma cells on bacterial nanocellulose scaffolds. *Artif. Cell. Nanomed. Biotechnol.* **2014**, *42*, 302–308. [CrossRef] [PubMed]

134. Abeer, M.M.; Amin, M.C.I.M.; Martin, C. A review of bacterial cellulose-based drug delivery systems: Their biochemistry, current approaches and future prospects. *J. Pharm. Pharmacol.* **2014**, *66*, 8. [CrossRef] [PubMed]

135. Silvestre, A.J.; Freire, C.S.; Neto, C.P. Do bacterial cellulose membranes have potential in drug-delivery systems? *Expert Opin. Drug Deliv.* **2014**, *11*, 1113–1124. [CrossRef] [PubMed]

136. Amin, M.C.I.M.; Ahmad, N.; Pandey, M.; Xin, C.J. Stimuli-responsive bacterial cellulose-g-poly(acrylic acid-co-acrylamide) hydrogels for oral controlled release drug delivery. *Drug Dev. Ind. Pharm.* **2014**, *40*, 1340–1349. [CrossRef] [PubMed]

137. Peres, M.d.F.S.; Nigoghossian, K.; Primo, F.L.; Saska, S.; Capote, T.S.d.O.; Caminaga, R.M.S.; Tedesco, A.C. Bacterial Cellulose Membranes as a Potential Drug Delivery System for Photodynamic Therapy of Skin Cancer. *J. Braz. Chem. Soc.* **2016**, *27*, 1949–1959. [CrossRef]

138. Hoffman, W.L.; Jump, A.A.; Ruggles, A.O.; Kelly, P.J. Antibodies bound to nitrocellulose in acidic buffers retain biological activity. *Electrophoresis* **1993**, *14*, 886–891. [CrossRef] [PubMed]

139. Mohammadi, H. Nanocomposite biomaterial mimicking aortic heart valve leaflet mechanical behaviour. *Proc. Inst. Mech. Eng. Part H J. Eng. Med.* **2011**, *225*, 718–722. [CrossRef] [PubMed]

140. Tronser, T.; Laromaine, A.; Roig, A.; Levkin, P.A. Bacterial Cellulose Promotes Long-Term Stemness of mESC. *ACS Appl. Mater. Interfaces* **2018**, *10*, 16260–16269. [CrossRef] [PubMed]

MDPI
St. Alban-Anlage 66
4052 Basel
Switzerland
Tel. +41 61 683 77 34
Fax +41 61 302 89 18
www.mdpi.com

Nanomaterials Editorial Office
E-mail: nanomaterials@mdpi.com
www.mdpi.com/journal/nanomaterials